REPRODUCTION, RACE, AND GENDER
IN PHILOSOPHY AND THE
EARLY LIFE SCIENCES

SUNY SERIES, PHILOSOPHY AND RACE
Robert Bernasconi and T. Denean Sharpley-Whiting, editors

REPRODUCTION, RACE, AND GENDER
IN PHILOSOPHY AND THE
EARLY LIFE SCIENCES

Edited by
SUSANNE LETTOW

Published by State University of New York Press, Albany

© 2014 State University of New York

All rights reserved

Printed in the United States of America

No part of this book may be used or reproduced in any manner whatsoever without written permission. No part of this book may be stored in a retrieval system or transmitted in any form or by any means including electronic, electrostatic, magnetic tape, mechanical, photocopying, recording, or otherwise without the prior permission in writing of the publisher.

For information, contact State University of New York Press, Albany, NY
www.sunypress.edu

Production by Dana Foote
Marketing by Michael Campochiaro

Library of Congress Cataloging-in-Publication Data

Reproduction, race, and gender in philosophy and the early life sciences / edited by Susanne Lettow.
    pages cm.—(SUNY series, Philosophy and race)
  Includes bibliographical references and index.
    ISBN 978-1-4384-4949-4 (hc alk. paper) 978-1-4384-4948-7 (pb alk. paper)
  1. Race—Philosophy 2. Human reproduction—Philosophy 3. Sex—Philosophy I. Lettow, Susanne, editor of compilation.
  HT1521.R456 2014
  305.8001—dc23
                                        2013005360

10 9 8 7 6 5 4 3 2 1

CONTENTS

Introduction
    Susanne Lettow   1

PART I. REPRODUCTION AND THE EARLY LIFE SCIENCES

1. Generation, Genealogy, and Time: The Concept of Reproduction from *Histoire naturelle* to *Naturphilosophie*
    Susanne Lettow   21

2. Organic Molecules, Parasites, *Urthiere*: The Controversial Nature of Spermatic Animals, 1749–1841
    Florence Vienne   45

3. The Scientific Construction of Gender and Generation in the German Late Enlightenment and in German Romantic *Naturphilosophie*
    Peter Hanns Reill   65

4. *Zeugung/Fortpflanzung*: Distinctions of Medium in the Discourse on Generation around 1800
    Jocelyn Holland   83

5. Treviranus' *Biology*: Generation, Degeneration, and the Boundaries of Life
    Joan Steigerwald   105

PART II. ARTICULATIONS OF RACE AND GENDER

6. Skin Color and the Origin of Physical Anthropology (1640–1850)
   Renato G. Mazzolini — 131

7. The Caucasian Slave Race: Beautiful Circassians and the Hybrid Origin of European Identity
   Sara Figal — 163

8. Analogy of Analogy: Animals and Slaves in Mary Wollstonecraft's Defense of Women's Rights
   Penelope Deutscher — 187

9. Reproducing Difference: Race and Heredity from a *longue durée* Perspective
   Staffan Müller-Wille — 217

10. Heredity and Hybridity in the Natural History of Kant, Girtanner, and Schelling during the 1790s
    Robert Bernasconi — 237

11. Sexual Polarity in Schelling and Hegel
    Alison Stone — 259

    About the Contributors — 283

    Index — 287

INTRODUCTION

SUSANNE LETTOW

In recent decades, the formation of the concept of race in the late eighteenth and early nineteenth centuries has attracted much scholarly interest particularly in the history of science, philosophy, and literary studies. At the same time, the naturalization of gender differences, which went hand in hand with the emerging life sciences, has been widely studied and criticized. However, the concept of race and the naturalized, scientific understanding of gender have rarely been studied in relation to each other, although their co-emergence is not just a question of simultaneity. At the end of the eighteenth century, the two ideas play a central role in the process of the temporalization of nature and the emergence of the life sciences. In particular, scientific understandings of race and gender are constituted and disputed within the debates on procreation, generation, and heredity that take place during the period. *Race* and *gender*[1] are thus closely connected to the new focus on diachronic processes of propagation and on long-term successions of individuals, which—in the second half of the eighteenth century—came to be articulated by the neologism *reproduction*.[2] However, the fact that concepts of race and gender co-emerged within the "procreation discourse" (Jocelyn Holland) of the late eighteenth century does not mean that they did so in parallel or homologous ways. On the contrary, connections between *race*, *gender*, and *reproduction*, which were of central importance for population politics later in the nineteenth century, were dispersed and unstable during the period.

The aim of this volume is to inquire into processes of the co-emergence of the concepts of race, gender, and reproduction in the decades around 1800—a period when all these concepts were in the making. To explore both continuities and discontinuities with subsequent biopolitical discourses, the volume examines specific configurations of biological and philosophical knowledge within their cultural and political contexts at the beginning of modernity. Philosophical discourse is a main focus of this inquiry because of its paramount influence in shaping the emerging field of the life sciences and because philosophy itself was reshaped in this process. The volume therefore not only contributes to a contextualist understanding of the life sciences, but also questions a modern, purified understanding of philosophical discourse by resituating it in the scientific and cultural contexts within which it emerged.

When, in the course of the eighteenth century, philosophers and scientists started to question the mechanistic understanding of nature and to identify a particular realm of nature where mechanic laws did not—or at least not sufficiently—apply, they envisaged "another world," as Charles Bonnet put it, "a new spectacle" (Abraham Trembley), or "a new nature" (Pierre-Louis Moreau de Maupertuis).[3] The capacities to reproduce, or self-generate and self-organize, soon became the crucial characteristics of the natural entities that belonged to this new realm of "living" nature. As Kant famously pointed out in the second part of the *Critique of Judgment,* an "organized being," which must be regarded as *Naturzweck* (natural purpose), is characterized by the fact that it is "of itself" cause and effect.[4] Kant gives the example of a tree, which is defined by its capacities of reproduction, growth, and regeneration or self-preservation.[5] In a certain sense, the whole enterprise of the temporalization of nature builds on the concept of reproduction, that is, on an understanding of procreation and generation that takes into account temporal change, the emergence of something new, and thus moves away from preformationism and the idea that, at creation, God formed the "germs" of all living beings, which from then on only have to be "unfolded." In short, *life*—the term that becomes the conceptual center of the new science named *biology*—is "that which produces, grows, and reproduces," as Michel Foucault has put it.[6]

In this epistemic context, ideas of race and gender underwent far-reaching changes as both became situated in and defined by the early life sciences. On the one hand, the scientific interest in processes of reproduction went hand in hand with disputes about the role of the sexes in propagation and culture in general. On the other hand, inquiries into human variation were increasingly concerned with processes of race-mixing, crossbreeding, and hereditary transmission of parental traits. Although "critical attention to early race theory" has for a long time focused on debates concerning monogenism and polygenism, Stefani Engelstein suggests that "the paradigm shift in reproductive theory . . . from preformation to epigenesis" is at least "equally important for the constitution of modern race discourse."[7] As scholars like Claudia Honegger and Londa Schiebinger have shown, it was just these "scientificalized" understandings of race and gender that played a prominent role in the period's political and ethical debates on equality and inequality, freedom and nonfreedom.[8]

The terms *race* and *sex* were in use long before the late eighteenth and early nineteenth century. In the case of *race*, François Bernier is said to have been the first to use the term as a way of distinguishing global populations in his *Nouvelle Division de la Terre, par les différentes Espèces ou Races d'hommes qui l'habitent* (New Division of the Earth, According to the Different Species or Races of Men who Inhabit It), published in 1684. However, as the title already suggests, Bernier did not distinguish between *races* and *species*. Rather, he used both terms in a broad sense to designate certain groups of human populations. This was also the case for Linnaeus and Buffon, who, if they used the term at all, did so unsystematically. As is now widely known, it was Immanuel Kant who first gave a definition of race that was to powerfully impact not only contemporary debates but also those of the subsequent centuries. Kant, as Robert Bernasconi puts it, not only gave a definition, but "set a direction for further inquiries."[9] Specifically, he did two things. First, he introduced the concept of race as a crucial term for his project of a genealogically oriented natural history, and thus as one that contributed to overcoming the classificatory, static forms of knowledge about nature, which Kant called *Naturbeschreibung* (natural description). In "the description of nature," he writes, "only the comparison of marks matters. What is here called *kind* [*Art*] is often only called *race* there,"[10] where "there" refers to *Naturgeschichte* (natural history) as Kant understands it. For Kant, then, the term *race* only functions within the new form of knowledge of nature, which deals with developments and genealogies instead of taxonomies. Second, and in accordance with this, Kant linked race to the concept of heredity. His definition is:

> Among the subspecies, i.e., the hereditary differences of the animals which belong to a single phylum, those which persistently preserve themselves in all transplantings (transpositions to other regions) over prolonged generations among themselves and which also always beget half-breed young in the mixing with other variations of the same phylum are called *races*.[11]

Interestingly, with this definition Kant also rearticulated the concept of heredity. As Staffan Müller-Wille and Hans Jörg Rheinberger have pointed out, "the way Kant set up the problem and the way he advanced a solution can be regarded as prototypical for the emergence of heredity."[12] Until then, *heredity* had mainly been used in a juridical sense with research and reflections on hereditary diseases as the exception.[13] With Kant's new articulation, however, the term came to designate intergenerational transmissions of bodily traits resulting from reproduction.

In a similar way, ideas of sex and gender underwent significant change in this period. In the German-speaking lands, the term *Geschlecht* had long been used in a genealogical sense, designating kinship groups like large families or dynasties. Toward the end of the eighteenth century, however, its meaning became more and more restricted, so that *Geschlecht* came to refer mainly to the sexes. The German dictionary *Zedlers Universal-Lexikon* from 1735, for example, only refers to the genealogical meaning but not to sexual difference in any natural-historical or biological sense—in clear contrast to dictionaries from the early nineteenth century.[14] Thus only when social change and the politics of equality—most obviously in women's activism and criticism during the French Revolution—undermined traditional gender arrangements and hierarchies, the bodily differences between women and men became the object of a new cultural concern. The emerging life sciences, physical anthropology, and in particular comparative anatomy thereby played an important role in the invention of modern gender polarizations and a naturalized, scientific understanding of gender that absorbed and legitimated cultural assumptions about difference and hierarchy.

However, a polarized, hierarchical understanding of gender developed only around 1800. As Peter Hanns Reill has shown in his comprehensive analysis of Enlightened vitalism,[15] mid-century theories of reproduction only occasionally referred to the different roles and functions of the sexes and their reproductive organs. When these were mentioned, as for example in Maupertuis's work, it was to stress the equal contribution of both sexes to the process of procreation and generation, contesting both the ovarist and the animalculist versions of preformationism, which argued for the preexistence of the individual in either the ovum or the sperm.[16] Even Johann Friedrich Blumenbach's concept of the *Bildungstrieb* (formative drive), which proved so important for the formation of a "science of life" because it allowed "the systematic level of the organization of an individual" to be linked with "the order of living nature,"[17] did not account for sexual difference. The *Bildungstrieb* in Blumenbach's sense was a homogeneous drive active in all living beings in the processes of structuring nutrition, generation, and regeneration. This view would change in the years to come. When Friedrich Wilhelm Joseph Schelling, in his philosophy of nature, adopted Blumenbach's concept, he added the idea that the drive "separates into opposing tendencies," male and female.[18] Like Kant's definition of race, this sexualization of the *Bildungstrieb* took on central importance for subsequent scientific theories and political-ethical discourse alike. However, as Peter Hanns Reill, Florence Vienne, and Alison Stone argue in this volume, the cultural process of the "polarization" and "differentiation" of the sexes was not a linear development; rather, during the period, both one-sex and two-sex models overlapped as they were favored by different authors.[19]

Though the invention and rearticulation of *race* and *gender* at the turn of the century was by no means a merely intellectual project, the fact that philosophy and the early life sciences were involved in this process cannot be underestimated. Indeed, the interplay of these fields of knowledge contributed much to the cultural significance that the concepts of race and gender would eventually gain. In the decades around 1800, philosophies of nature, including German *Naturphilosophie*, were closely tied to the scientific debates and developments of the day. As historian of science Dietrich von Engelhardt notes,

> All the key philosophers of the time thought about the relation of man to nature and the possibilities of knowledge about nature to an extent that has never been matched since then. In addition, though not to the same extent, naturalists around 1800 grappled directly with the conceptions of *Naturphilosophie*.[20]

At this point, neither the life sciences nor philosophy had clear disciplinary demarcations; they were, precisely, emerging fields of knowledge. In certain respects it seems that they shaped one another and benefited mutually from the cultural credit each gained in this period. Speaking of *Naturphilosophie* was thus "perceived as offering natural history a rise in status from a mere appendage of the medical faculty to full membership alongside mathematics, philology, and physics in a higher philosophical faculty."[21] The avoidance of philosophy by natural scientists only arose later in the nineteenth century, when biology had become established as a natural science. In contrast to biology, philosophy had been institutionalized as a distinct field of knowledge for centuries. Yet this institutionalization, and accordingly the cultural understanding of philosophy, also changed significantly around 1800. In the medieval arrangement of disciplines that prevailed in higher education in Germany until the end of the eighteenth century, philosophy was regarded as a mere preliminary "to the higher studies in one of the credentialed professions, theology, law or medicine."[22] This subaltern position of the philosophical faculty was, however, highly contested in the disputes of the 1790s that led to the Prussian academic reform. This reform, in which most of the protagonists of what became known as *classical* German philosophy were involved, transformed the philosophical faculty "into a full-fledged higher faculty, claiming to teach the most advanced subjects, and with autonomy from the restrictions formerly imposed by the theologians."[23] Thus, in the decades around 1800, both philosophy and the life sciences were in the making and—in this process of emergence—closely interrelated. The disciplinary borders that were established in the course of the nineteenth century and the gulf that was later thought to separate the natural sciences from the humanities and social sciences did not yet exist.

Neither did the idea of nature as entirely separated from the social or the cultural. With respect to the emerging concepts of race, gender, and reproduction, which were to become main points of reference for the biopolitical discourses of the nineteenth, twentieth, and even twenty-first centuries,[24] this meant that the way in which scientific and political-ethical meanings of these concepts intersected in the period differed considerably from their later articulations. As Lorraine Daston has argued, "naturalization" as a discursive process in which social, political, and cultural ideas are projected into nature was by no means new at the end of the eighteenth century. On the contrary, the "moral authority of nature" could be and was interpellated in various ways, as Daston and Fernando Vidal show by using examples from different periods and societies.[25] However, "where conceptions of nature diverge, so do the strategies . . . of naturalization,"[26] and this was certainly the case in the late eighteenth century. According to Emma Spary, the new mode of naturalization can be traced back to Buffon. At least "the appeal to a 'nature' existing beyond society, but forming the bedrock of social interactions and moral aesthetic standards, was what many readers particularly perceived in Buffon, even if there was little consensus on his political or religious views."[27] A specific form of naturalization was thus appearing on the horizon, to which later forms of "scientific racism," biological articulations of gender differences and various forms of population policy could refer, even if the configurations of the late eighteenth and early nineteenth century clearly differ from the later forms of biological determinism that presuppose a sharp distinction between the social and the natural.

From the mid-nineteenth century onward, the idea that sex and race are biological givens indicating cultural and social status and that these concepts—along with those of the child, the criminal, and the insane—can be treated as analogies, became common currency, as the relentless repetition of these ideas shows.[28] However, in the period between 1750 and 1830, which is the focus of this volume, a scientific understanding of race, a dichotomized understanding of sexual difference, and analogies between concepts of race and gender were relatively new, contested, and unstable. Often, scholars did not explicitly acknowledge the mutual implications of their research on gender or race because the two strands of research were conducted separately. In particular, research on race mainly focused on men while sexual difference was seen as part of the struggles over European bourgeois gender arrangements.[29] The studies collected here, nevertheless, show how concepts of race and gender emerged within the same epistemic and political-cultural context and that they became systematically interlinked via reflections on reproduction. At the same time, contributions show that conceptual and political connections were loose because the concepts at stake were in the making and circulated in different fields of knowledge and because they coexisted with older views. Therefore, a critical analysis of gender and race discourses has to address the "multiplicity" of concepts.[30]

Thus, the concept of race was not only relevant in its genealogical meaning, linking it to theories of crossbreeding and heredity, but also borrowed from the long tradition of scientific inquiry into the causes of skin color. The invention of "whiteness," as Renato Mazzolini shows in this volume, is closely connected to this kind of research as well as to ideas of beauty and ugliness within the widely circulating travel literature of the period. At the same time, the aesthetic articulation of racial hierarchies that culminated in the invention of the political phantasma of a beautiful, superior "Caucasian" race is highly gendered. As Sara Figal's contribution shows, the fascination with the Caucasus and the Caucasian people was not only due to a reappropriation of ancient mythology and the biblical story of the landing of Noah's Ark on Mount Ararat, but above all draws on eroticized reports of the beauty of Circassian women, especially female slaves. Here, the phantasmagoric dimensions of concepts of race and their articulation with erotic desire and gender politics are obvious. It also becomes clear in how closely the invention of racial hierarchies is intertwined with the issue of slavery. However, it is not only the economics and politics of slavery and abolitionism that fueled racial discourse. In the eighteenth century, "slavery" had become a general signifier for oppression, domination, and nonfreedom.[31] Slavery thus also functioned as a political metaphor, which, for example in the rhetoric of Mary Wollstonecraft,[32] evoked analogies between women's oppression and the oppression of slaves. Despite the attempts to define *race*—as in Kant's work—the concept thus frays and blurs. It crosses different fields of knowledge and discourses and simultaneously connects them.

A similar process can also be seen in ideas about propagation. The different terms that designate these processes—*procreation, generation, reproduction, regeneration*, or *degeneration*—are all used in new, previously unthought-of ways, and their meanings shift further as they are introduced into new domains. In the wake of new scientific, social, and political uncertainties about the meanings, status, and cultural relevance of lineage and kinship relations, a kind of genealogical anxiety manifests itself in the decades around 1800.

In these years, relations of reproduction—the social relations and arrangements that regulate the "making" of children and conceptualizations of reproduction—underwent far-reaching change. In particular, the success of epigenesis has to be seen in this context. It was, at least partly, due to the fact that epigenesis was compatible with the new ideas and practices of free-choice marriage and romantic love. In contrast to preformationism, according to which the choice of a specific "generative partner" was insignificant, within an "epigenetic framework . . . there is always an 'alternative,'" opening up the possibility of a free choice for "those partners who desire to 'mix.'"[33] The theoretical debates on procreation and generation thus clearly contributed to new understandings of marriage, love, and gender relations in the late eighteenth- and early nineteenth centuries.[34] However, within these theories, different stances on the function and status of the sexes coexisted so that these theories seem to have been a medium within which divergent ideas about gender and gender equality were contested.

To sum up, the philosophical and scientific debates that contributed to the formation of the early life sciences evoke and are situated in various social and cultural contexts—contexts that are to some extent interconnected, but cannot be homogenized. From the present day point of view, studying the early life sciences and their political-ethical dimensions and contexts, and especially studying concepts of race, gender, and reproduction, is of particular interest because the historical distance that separates us from these configurations allows to trace out both continuities and discontinuities. The contributions to this volume indicate that the texts and debates of the late eighteenth and early nineteenth centuries elude the idea of biopolitics as a unified paradigm that emerged during this period and which has shaped modernity since then. Moreover, these analyses contribute to a more complex understanding of epistemic processes that fed the formation of biology and biopolitical discourse.

The first section of the book focuses on the importance of theories and concepts of procreation, generation, and reproduction for the emergence of the life sciences. Contributions of this section scrutinize the various articulations and the cultural contexts in which they circulated. Susanne Lettow opens this section with an analysis of the emergence and circulation of the concept of reproduction from the mid-eighteenth century to German *Naturphilosophie*. Lettow argues that up to the end of the century, the concept of reproduction became more and more invested with cosmological ideas about continuity and individuals' belonging to supra-individual entities like the species, sex, or race. Whereas authors like LaMettrie, Maupertuis, and Buffon or Blumenbach, Herder, and Kielmeyer dealt with heterogeneous models of temporality that at least partly acknowledged temporal change and an open future, in the *Naturphilosophie* of Schelling, Görres, and Hegel, the understanding of time became restricted, homogenized, and mythologized. This, in Lettow's view, contributed to the emergence of a biopolitical gaze.

Florence Vienne's chapter focuses on research on spermatic animals between 1749 and 1805; that is, between the publication of Buffon's *Histoire naturelle* and Lorenz Oken's *Die Zeugung*. By comparing different understandings of the reproductive substance as organic molecules, parasites, or *Urthiere*, Vienne shows how the different theories were all permeated by assumptions about gender relations. She argues that the still-current idea of sperm as characterized by motility and vitality originates from these debates, particularly in the dualistic construction of gender that obsessed Romantic *Naturphilosophen* like Oken.

Both Lettow's and Vienne's accounts accord with Peter Hanns Reill's claim that, by the end of the century, scientific constructions of gender and a hierarchical, polarized understanding of gender difference increasingly replaced earlier conceptualizations of gender that highlighted ambiguity and the interaction between differences. Reill's chapter analyzes the articulations of gender difference in the theories of generation formulated by Wilhelm von Humboldt, Lorenz Oken, and Carl Gustav Carus. Humboldt, who was fascinated by androgyny and "strong" women in both the mental and physical sense, conceived of two forces—both active although different—that worked together in generation. But the Romantic *Naturphilosophen* some years later transformed difference into a hierarchical polarization. Reill argues that the "blurring of boundaries and the transmutation of gender categories" of the late Enlightenment became more and more closely identified with the turmoil and uncertainties resulting from the French Revolution. The *Naturphilosophen* yearned instead for "order, clarity, and hierarchy." The gender constructions of Oken and Carus are particularly striking examples of this obsession with clear separations between the sexes and with the devaluation of women, who came to be equated with "the fish" or "the plant," and thus with the lower principles of living nature.

All these debates were part of a broad cultural concern with procreation and generation. Not only did scientific ideas about the reproductive process and the materials involved in it change, but the concepts of procreation and generation themselves were deployed in new ways. Jocelyn Holland inquires into the new meanings acquired by the terms *Zeugung* (procreation) and *Fortpflanzung* (generation) within the German intellectual community. Although the concepts were not clearly distinguished and often used interchangeably, *Fortpflanzung* and *Zeugung* are, as Holland shows, "indebted to different points of departure and serve different purposes." Whereas *Zeugung* mainly refers to singular acts of producing a new individual, *Fortpflanzung* is articulated with notions of temporal expansion and duration. Pointing to Goethe, Herder, and Fichte, Holland argues that *Fortpflanzung* tends to be construed as a virtual medium within which the real acts of procreation happen. This understanding of *Fortpflanzung* is not limited to the organic realm but extends into, for example, Ritter's theories of chemistry and acoustics. Here, the "virtualization" of propagation becomes a central discursive operation. The debates Holland reconstructs are dispersed and relate to different fields of knowledge, but appear to have influenced nineteenth-century biology. In her chapter on Gottfried Reinhold Treviranus, one of the authors who invented the term *biology*, Joan Steigerwald shows how the initial outlines of this new science resulted from an amalgamation of earlier debates and developments. Treviranus, Steigerwald argues, was neither an original thinker nor an innovative researcher, but his "mixture of approaches" made the six volumes of *Biology, or Philosophy of Living Nature for Natural Researchers and Physicians* (Biologie, oder Philosophie der lebenden Natur für Naturforscher und Ärzte), which appeared between 1814 and 1822, an "ambiguous prototype for a new science of life." The notion of "bios," or "life," which was the conceptual core absorbing these different approaches, was as Steigerwald points out, shaped by a peculiar interest in the boundaries and border zones of life, especially in those living beings that Treviranus called *animal-plants* or *plant-animals*. This interest led him to highlight not only "the regularity and organization" of living beings but also their contingent and deviant aspects, thus not only manners of generation but also of degeneration. Although

Treviranus's sketch of biology dismissed the political-ethical dimensions that helped to shape this field of knowledge, he incorporated much from the earlier debates on reproduction, gender, and race.

The second part of the volume shows how concepts of race and gender were articulated within the context of the early life sciences, how they resonated with each other, and how they circulated in the scientific, philosophical, and political-ethical debates of the period. Although gender and race were linked through ideas of crossbreeding and heredity, the meaning of these concepts was by no means confined to the life sciences. Moreover, modern gender and race discourses have to be understood as amalgamations of scientific, political, aesthetic, and phantasmatic elements.

Renato Mazzolini reconstructs the history of skin-color research from the mid-seventeenth to the mid-nineteenth century. He argues that the classifications built on this research, which focused solely on "black" skin, preceded concepts of race and modern racism. They were deeply influenced by the perception of Africans that emerged in the context of the system of color-based slavery introduced by the Europeans in the early sixteenth century. However, the political implications of these classificatory differences were highly disputed. Whereas many earlier researchers contested justifications of the subjugation of Africans by presenting "natural" explanations for differences in skin color, toward the turn of the eighteenth to the nineteenth century, these differences were increasingly equated with aesthetic and intellectual distinctions and hierarchies. In this period, too, the idea that Europeans also have a skin color—white—first emerged.

Sara Figal's contribution shows that the invention of whiteness and the European Caucasian race also was a highly imaginative and phantasmagoric process. Figal examines the figure of the female Circassian slave, which stood at the center of race theories, specifically in narratives about the superior Caucasian race. As Figal argues, this figure points to the highly ambivalent, hybrid status of white European identity because it locates the origin of that identity outside the cultures considered "civilized" and outside of orthodox Christendom; in other words, in a world that, according to travel writers of the time, was populated by "animist heathens with vestiges of Christian and Muslim influence" (Figal, in this volume). In addition, the fascination that captured writers such as Bernier or John Chardin was highly invested with erotic desire and Orientalist fantasies about the "female harem slave." However, it was not only "eroticized exoticism" that made the beautiful Circassian an icon of racial theories, but also proto-eugenic fantasies and reflections on crossbreeding; that is, on the improvement of blood and beauty through reproducing with Circassian or Georgian women.

A different entanglement of gender politics and racial discourse is analyzed by Penelope Deutscher, who scrutinizes Mary Wollstonecraft's feminist rhetoric. Deutscher points out how Wollstonecraft's *Vindication of the Rights of Women,* published in 1792 at the height of parliamentary debates on slavery, is pervaded by a set of analogies among slaves, animals, and women. Some of these analogies, especially the analogy between the slave and the animal as subjected to the cruel behavior of European men, had already circulated in other feminist and abolitionist writings like those of Catherine Macaulay or Olaudah Equiano. Only in Wollstonecraft, however, are these analogies entirely deprived of literal meaning, unfolding instead a complex interplay of "analogies of analogies." Although Wollstonecraft seems, at least partly, to be conscious of the metaphorical use she makes of figures like the animal or the slave to decry women's oppression, she cannot reflect on the "rhetorical profit" she gains from these analogies. As Deutscher argues, to qualify the subordination of women as abhorrent because it degrades them to the status of slaves also means implicitly to reiterate the subordination of slaves, even if Wollstonecraft was highly critical of slavery.

From the perspective of the history of science, Kant's definition of race through heredity surely was a crucial moment in the modern history of the concept. Staffan Müller-Wille, however, in his chapter on race, heredity, and disease introduces a *longue durée* perspective on these interrelated phenomena. He argues that, far from being an essentialist concept built on the idea of an unchangeable set of fixed traits, the modern concept of race arose from thinking about hereditary variation so that "deviation" is part of its core meaning. This idea of deviation, Müller-Wille writes, can be traced back to early-modern research and reflections on hereditary diseases. Thus, although the concepts of race and heredity surfaced in the years around 1800, their history extends in both directions: into the past and into our own time, where racial categories are reappearing in the context of genomics and biomedicine.

Robert Bernasconi also addresses the relationship of race and heredity. He focuses on the debates of the 1790s with particular attention to the contributions of Kant, Girtanner, and Schelling. He starts from the seeming paradox that Kant's concept of race, widely adopted among philosophers and naturalists by the end of the century, was linked to a preformationist concept, namely that of germs. Bernasconi shows that Kant's notion of germs was complex and that he developed a specific position called *generic preformationism* that allowed him to reconcile elements from epigenesis and preformationism. The epigenetic orientation Kant laid out in the *Critique of*

Judgment thus did not contradict his earlier position formulated in the race essays because Kant never fully abandoned the notion of germs. The wide circulation of Kant's concept of race, largely through the writings of Girtanner and Schelling, underlines Bernasconi's argument that the *Critique of Judgment* and the adaptation of Blumenbach's *Bildungstrieb* were anything but a "graveyard" for the concept of race.

Schelling, however, did not only contribute to the circulation of the Kantian concept of race. Like Hegel—although within a different philosophical framework—he proposed a notion of reproduction marked by sexual polarity. As Alison Stone shows, Schelling deploys the idea of two opposite forces, production and inhibition, that work together in reproduction and are geared toward overcoming sexual difference for the sake of the species' unity. In Hegel, however, the main conceptual distinction is between concept and matter. Like Schelling, Hegel regards the opposition between the sexes as a prerequisite for the generation of a "third," and thus as part of an infinite dynamics, and he overtly sexualizes the distinction between concept and matter that is supposed to structure the whole of nature. This distinction also plays a major role in Hegel's theory of race. The ideas on sexual and racial difference are thus clearly connected but, Stone argues, hold a differing status in his philosophy. While the notion of sexual polarity aims (unsuccessfully) to overcome the distinction of concept and matter, racial difference "reflects the fact that the opposition . . . cannot be overcome at the level of the natural soul" (Stone, in this volume). Racial differences thus seem to be even more static than sexual difference.

Taken together, the chapters in this volume show that within the epistemic debates of the early life sciences and philosophies of nature, concepts of race and gender played a crucial role. Closely connected to ideas of crossbreeding, reproduction, and heredity, concepts of race and gender clearly resonate with each other, although meanings of race and gender disperse into various political-ethical discourses. Instead of a series of analogies among "women," "lower races," "savages," and "criminals" that were drawn in nineteenth-century discourse with reference to evolutionary biology,[35] we find relations of resonance. This means that concepts of race and gender correlate but are rarely parallelized or treated as analogues. To take such differences between epistemic-political constellations into account allows one to acknowledge both similarities and dissimilarities and thereby contributes to a *longue durée* analysis of the intersections of race and gender ideologies and the specific cultural status of biological knowledge in modernity. Particularly with regard to recent (re-)articulations of race and gender in the context of biomedicine and the neurosciences, such a *longue durée* perspective can contribute to a better understanding of the cultural, political, and scientific stability of concepts of race and gender.

The initial idea for this volume arose from two workshops that were held in Vienna at the Institut für die Wissenschaften vom Menschen (IWM)/Institute for Human Sciences in the context of my research project on the "The symbolic power of biology: Articulations of biological knowledge in *Naturphilosophie* around 1800." I am grateful to Cornelia Klinger, who invited me with this project to the IWM, which provided a wonderful working atmosphere for years. She has followed this project with friendly advice and support in many respects. Beyond the IWM, I was happy to have the opportunity to share my thoughts with Nick Hopwood and the research group "From Generation to Reproduction" during my stay at the Department for History and Philosophy of Science and Clare Hall College at Cambridge University in early 2011. In Germany, the research network "Economies of Reproduction 1750–2010," funded by the German Research Foundation, provided an inspiring working context. In particular, I thank Florence Vienne for the many discussions about reproduction, temporality, and the early life sciences we have had in recent years. I am also thankful to her and to Staffan Müller-Wille for comments on an earlier version of the introduction, and to Kate Sturge and Justin Rainey for copyediting and proofreading. In financial respects, the volume and the research from which it originated have been made possible by a grant from the Austrian Science Fund (FWF). I thank the FWF for the gratuitous support.

NOTES

1. Although the term *gender* only acquired its contemporary, critical meaning in the context of feminist theory in the late twentieth century, I prefer—where context allows—to use the term *gender* instead of *sex* because it highlights the sociocultural implications of all concepts and ideas of sexual difference. After all, the whole sex-gender-distinction, which presupposes a clear demarcation between nature and culture, or the biological and the social does not apply to eighteenth- and early nineteenth-century materials, so that sex is always already gender. Cf. Ludmilla Jordanova, "Sex and Gender," in *Inventing Human Science: Eighteenth-Century Domains*, ed. Christopher Fox, Roy Porter, and Robert Wokler (Berkeley, Los Angeles, London: University of California Press, 1995), 152–183.

2. I use the term *reproduction* where I refer to the neologism that was coined in the eighteenth century. *Procreation, generation,* and *propagation* are used in a broad rather unspecific way although distinctions between these terms were also drawn and debated in the period (see Jocelyn Holland, in this volume).
3. Cited by Giulio Barsanti, "Les phénomènes 'étranges' et 'paradoxaux' aux origines de la première revolution biologique (1740–1810)," in *Vitalisms from Haller to Cell Theory,* ed. Guido Cimino and François Duchesneau (Florence: Leo S. Olschki, 1997), 67–82, p. 67.
4. Immanuel Kant, *Critique of the Power of Judgment,* vol. 5 of *The Cambridge Edition of the Works of Immanuel Kant,* ed. Paul Guyer, trans. Paul Guyer and Edward Mathews (Cambridge: Cambridge University Press, 2000), 370, 243.
5. Ibid.
6. Michel Foucault, *The Order of Things: An Archaeology of the Human Sciences* (London: Routledge, 1991), 232.
7. Stefani Engelstein, *Anxious Anatomy: The Conception of the Human Form in Literary and Naturalist Discourse* (Albany: State University of New York Press, 2008), 226.
8. Cf. Londa Schiebinger: *Nature's Body: Gender in the Making of Modern Science* (New Brunswick: Rutgers University Press, 2004); Claudia Honegger: *Die Ordnung der Geschlechter: Die Wissenschaften vom Leben und das Weib* (Frankfurt am Main and New York: Campus, 1991).
9. Robert Bernasconi, introduction to *Bernier, Linnaeus and Maupertuis: Concepts of Race in the Eighteenth Century* (Bristol: Thoemmes Press, 2001), vol. 1:vii–xiii.
10. Immanuel Kant, "Determination of the Concept of a Human Race," in *The Cambridge Edition of the Works of Immanuel Kant,* ed. Paul Guyer, trans. Paul Guyer and Edward Mathews (Cambridge: Cambridge University Press, 2000), vol. 8:100, 153.
11. Immanuel Kant, "Of the Different Races of Human Beings," in *Anthropology, History, and Education,* ed. Günther Zöller and Robert Louden, trans. Mary Gregor, *The Cambridge Edition of the Works of Immanuel Kant,* ed. Paul Guyer and Alan W. Wood (Cambridge: Cambridge University Press, 2007), vol. 2:85.
12. Staffan Müller-Wille and Hans-Jörg Rheinberger, "Heredity—the Formation of an Epistemic Space," in *Heredity Produced: At the Crossroads of Biology, Politics, and Culture, 1500–1870,* ed. Staffan Müller-Wille and Hans-Jörg Rheinberger (Cambridge, MA: MIT Press, 2007), 3–34, 18.

13. See Staffan Müller-Wille in this volume.
14. Cf. Ute Frevert, *"Mann und Weib, und Weib und Mann." Geschlechter-Differenzen in der Moderne* (München: C. H. Beck, 1995).
15. See Peter Hanns Reill: *Vitalizing Nature in the Enlightenment* (Berkeley, Los Angeles, London: University of California Press, 2005).
16. On preformationist theories, criticism of them, and the ideas of gender difference evoked in the debates, see Florence Vienne in this volume. Cf.: Shirley Roe, *Matter, Life and Generation. Eighteenth-Century Embryology and the Haller-Wolff-Debate* (Cambridge: Cambridge University Press, 1981); Clara Pinto-Correia, *The Ovary of Eve: Egg, and Sperm and Preformation* (Chicago: University of Chicago Press, 1997).
17. Peter McLaughlin, "Blumenbach und der Bildungstrieb: Zum Verhältnis von epigenetischer Embryologie und typologischem Artbegriff," *Medizinhistorisches Journal* 17 (1982):357–372, 358. My translation.
18. Friedrich Wilhelm Joseph Schelling, *First Outline of a System of the Philosophy of Nature*, trans. Keith R. Peterson (Albany: State University of New York Press, 2004), 36.
19. Thomas Laqueur has argued that during the eighteenth century the older one-sex model, according to which female bodies were conceived as default or minor versions of the male body was replaced by a two-sex model that highlights sexual dimorphism. His historical analysis, however, has been criticized for being too one-dimensional. Cf. for example Katherine Park, "Cadden, Laqueur and the 'One-Sex-Body,'" *Medieval Feminist Forum* 46, 1 (2010):96–100.
20. Dietrich von Engelhardt, *Hegel und die Chemie: Studie zu Philosophie und Wissenschaft um 1800* (Wiesbaden: Guido Pressler, 1976), 1. My translation.
21. Nicholas Jardine, "*Naturphilosophie* and the Kingdoms of Nature," in *Cultures of Natural History*, ed. Nicholas Jardine, James A. Secord, and Emma Spary (Cambridge: Cambridge University Press, 1996), 230–245, 243–244.
22. Randall Collins, "The Transformation of Philosophy," in *The Rise of the Social Sciences and the Formation of Modernity: Conceptual Change in Context, 1750–1850*, ed. Björn Wittrock, Johan Heilbron, and Lars Magnusson (Dordrecht: Kluwer, 1998), 141–161, 147.
23. Randall Collins, "The Transformation of Philosophy," 150. The most prominent contribution to this development was probably Kant's *Der Streit der Fakultäten* (The dispute of faculties), published in 1798.

24. On the persistence of race categories, see Müller-Wille in this volume; also Sander Gilman, *Diseases and Diagnoses: The Second Age of Biology* (New Brunswick: Transaction, 2010); Nicole Karafyllis and Gotlind Ulshöfer, eds., *Sexualized Brains: Scientific Modeling of Emotional Intelligence from a Cultural Perspective* (Cambridge, MA: MIT Press, 2010).
25. See Lorraine Daston and Fernando Vidal, "Introduction: Doing What Comes Naturally," in *The Moral Authority of Nature*, ed. Lorraine Daston and Fernando Vidal (Chicago: University of Chicago Press, 2003), 1–20.
26. Lorraine Daston, "The Naturalized Female Intellect," *Science in Context* 5 (1992):209–235, 210.
27. Emma Spary, "The 'Nature' of Enlightenment," in *The Sciences in Enlightened Europe*, ed. William Clark, Jan Golinksi, and Simon Schaffer (Chicago: University of Chicago Press, 1999), 272–304, 296.
28. See Nancy Leys Stepan, "Race and Gender: The Role of Analogy in Science," *Isis* 287 (1986):261–277.
29. Londa Schiebinger, *Nature's Body*, chapter 5 "Theories of Gender and Race," 143–183.
30. As Roxann Wheeler argues, "an awareness of multiplicity is based on the premise that in historical terms, ideologies and practices do not disappear; rather they coexist with new ways of thinking and living, are revised partially to fit new conditions, or 'go underground' for a while and resurface later." In addition, "a theory of multiplicity also acknowledges the various components of racial ideology. . . . A focus on the elasticity of race . . . is not to say that it was unconstrained, but its articulation was far more heterogeneous than is usually conceded today." See Roxann Wheeler, *The Complexion of Race: Categories of Difference in Eighteenth-Century British Culture* (Philadelphia: University of Pennsylvania Press, 2000), 39.
31. See Susan Buck-Morss, *Hegel, Haiti, and Universal History* (University of Pittsburgh Press, 2009).
32. See Penelope Deutscher in this volume.
33. Helmut Müller-Sievers, *Self-Generation: Biology, Philosophy and Literature Around 1800* (Stanford: Stanford University Press, 1997), 30.
34. For an analysis of the cultural concern with love and marriage in the period see Adrian Daub, *Uncivil Unions: The Metaphysics of Marriage in German Idealism and Romanticism* (Chicago: University of Chicago Press, 2012).
35. See Nancy Leys Stepan, "Race and Gender."

Part I

REPRODUCTION AND THE
EARLY LIFE SCIENCES

1

GENERATION, GENEALOGY, AND TIME

The Concept of Reproduction from *Histoire naturelle* to *Naturphilosophie*

SUSANNE LETTOW

While the concepts of procreation and generation, or *Fortpflanzung* and *Zeugung*, have been used for centuries to designate processes of the making and emergence of new human or animal beings, "reproduction" is a relatively new concept. As François Jacob stated, until the end of the eighteenth century

> "living beings did not reproduce; they were engendered. Generation was always the result of a creation which, at some stage or other, required direct intervention by divine forces. [. . .] The generation of every plant and every animal was, to some degree, a unique, isolated event, independent of any other creation, rather like the production of a work of art by man."[1]

With the idea of reproduction, attention shifted from singular acts and their circumstances to processes of propagation on a supra-individual level. As a consequence, "the reproduction of organic beings, in contrast to their mere production, became recognizable as a domain governed by laws of its own."[2] This, however, did not only enable new ways of scientific inquiry but was also a symptom of a transformation in genealogical thinking in the context of a broader process of social and cultural change, as I argue in this chapter.

In particular, the concept of reproduction—because of its temporal index—facilitated questions about continuity and discontinuity that were asked and answered in new ways in the late eighteenth and early nineteenth

21

centuries. Ideas of repetition, return, and eternity clearly belong to its semantic horizon as these were evoked in different uses of the concept by naturalists and philosophers. In a period when, as Reinhard Koselleck and others have argued, the cultural understanding of time underwent radical change and a new sense for the historicity of all being developed, questions of temporality discussed in natural history clearly had a huge cultural impact. The whole process of the temporalization of nature and knowledge about nature, in which the static classifications of natural history were increasingly criticized and replaced by ordering patterns that allowed for grasping the temporal dynamic of "living beings," went hand in hand with sociocultural processes, in which "time itself gain[ed] a historical quality."[3] The experience of the political, social, and cultural transformations of the late eighteenth century significantly fostered the perception of an increased acceleration and "the development of an open future."[4] In this context, genealogical thinking about diachronic and synchronic forms of belonging, or about lineage, descent, and kinship also underwent far-reaching change. The meaning of belonging to a "family," a "Geschlecht,"[5] a "sex," or "humanity" at large became increasingly uncertain when the social bonds that constituted these social formations wavered. By the end of the eighteenth century, kinship, family structures, and gender relations were reorganized through the emerging nation-states and population politics, and the expansion of geographical and cultural horizons through the European colonial expansion led to the questioning and reestablishment of hierarchical ordering patterns of human "kinship."

Against this backdrop, the co-emergence or reformulation of concepts of race, heredity, species, and sex in the process of the temporalization of nature can be understood as an attempt to reconfigure genealogy and belonging to social and natural orders. In one way or another, all these concepts articulate bonds of individuals with other humans or with nonhuman nature, and their more or less simultaneous appearance indicates the explosiveness of questions of the belonging of individuals to supra-individual entities. The concept of reproduction, however, seems to be at the center of this reinvention of genealogy because it articulates the concepts of race, heredity, and sex as well as the more abstract concepts of life, organism, and development. Moreover, the concept of reproduction clearly rivals the idea of Creation. As Ludmilla Jordanova has argued with respect to Lamarck's distinction between creation and production, these terms "not only stood at the very heart of the Christian tradition, which makes fundamental distinctions between begetting, that is, re-production, and creating, they also expressed concerns so basic to human existence that it is hard to articulate them without banality."[6] These terms express fundamental concerns about human agency and its limits—about the making and the being-made of humans. As such, they touch on questions that, since time immemorial, have been addressed by mythologies, religions, and philosophies and have provoked scientific curiosity. The theories of nature and generation formulated in the late eighteenth and early nineteenth centuries by philosophers and naturalists alike can thus be understood as "offering versions of creation myths."[7] "When natural scientists and philosophers theorized human and nonhuman nature they also attempted to understand," as Jordanova has put it, "the palpable affinities between living beings, especially human beings, as well as between God and nature, nature and nature's laws. Representing and conceptualizing such kinship in terms of visible similarities was a complex enterprise, one that touched not only a sense of cosmological order but also the most intimate identities of all concerned."[8]

In what follows, I discuss the emergence and the circulation of the concept of reproduction in the second half of the eighteenth century and the early nineteenth century. After a brief outline of the historical semantics of the concepts of production, generation, and reproduction, I present three different conceptual constellations in which the concept of reproduction surfaces and circulates. Rather than advance claims about continuity and discontinuity, I want to highlight differences between arrangements of semantic elements. The first constellation comprises the mid-century interventions of Julien Offray de La Mettrie, Pierre-Louis Moreau de Maupertuis, Georges-Louis Leclerc, Comte de Buffon, who all refer to the experiments of Abraham Trembley and his discovery of the ability of the sweet water polyp to regenerate and transform into two new polyps when cut into pieces. Second, I refer to the German debates on the characteristics of living nature in the late eighteenth century with particular attention to the contributions of Johann Friedrich Blumenbach, Johann Gottfried Herder, and Carl Friedrich Kielmeyer. The third constellation, then, comprises the philosophies of nature of Friedrich Wilhelm Joseph Schelling, Joseph Görres, and Georg Friedrich Wilhelm Hegel. It becomes clear that articulations of reproduction clearly differ between the mid-eighteenth century and the early nineteenth century, as well as between French debates on generation and the debates among naturalists and philosophers of nature in the German lands. The main differences occur with respect to the relation of the individual to supra-individual entities, to the role of the sexes, and to temporality. Despite these differences, however, it is striking that all these attempts to address processes of generation from a supra-individual perspective—from Buffon's *Histoire Naturelle* onward—are highly invested with cosmological ideas. However, I argue in the concluding remarks that the transformation of genealogical thinking that is constituted through concepts of sex, race, and heredity and new understandings of diachronic and synchronic belonging was not only closely intertwined with the cultural invention of new myths of origin that responded to experiences of uncertainty. The intellectual concern with reproduction, genealogy, and the belonging of individuals to supra-individual entities like the species, the sex, or the race also contributed to the emergence of a biopolitical gaze that addressed humans as subjugated to these new biosocial entities and to a new understanding of kinship relations.

GENERATION, PRODUCTION, REPRODUCTION: HISTORICAL-SEMANTIC SHIFTS

In contrast to *reproduction*, the terms *generation* and *production* can be traced back to earlier centuries. In French, *generation* was already used in the thirteenth century to designate the "production of a being by his parents."[9] The introduction of the term *productio* into Latin is attributed to the Stoic philosopher Proclus, who translated the Greek *paragogé* to designate the emanation of the One. In the Stoa, nature—as a lower stage of being—was conceived as "produced" and "producing" at the same time, that is, as having emanated from the highest being and having the ability to emanate something else. Later uses sometimes merge *production* and *creation* and sometimes draw distinctions between these two forms of making, God's *creatio ex nihilo* and *productio*, where the latter means making something out of something else already existing.[10] From the seventeenth century onward, however, *production* is used in a broad sense, so that it surfaces as an equivalent to a series of cognate terms, all designating forms of making: it is the "actio per quam aliquid fit sive generando sive creando et procreando sive efficiendo et producendo."[11] The use of the term *production* by naturalists and philosophers in the eighteenth century is close to this broad usage as well as to the early understanding of production as the activity of nature. The *Encyclopédie* (1765), for example, explains production as "all phenomena of nature of which the existence of a plant, a tree or whatever substance is the aim."[12] Accordingly, the introduction of the term *reproduction* links views on generation to ideas about natural processes in general, or to nature's productivity.

An indicator for the novelty of the term *reproduction* in the eighteenth century is the lack of entries in dictionaries like the *Dictionnaire de l'Académie Françoise* (1778) or the German *Zedlers Universal-Lexikon* (1732–1754). The latter, for example, has entries on *Zeugung* and *Generatio*, and terms such as *Erzeugung* and *Fortpflanzung* are frequently used.[13] Even the entry

"Wiederherstellung, Wiedererzeugung" refers neither to the term *reproduction* nor to Abraham Trembley's famous experiments on the sweet water polyp. In contrast, the *Encyclopédie* (1765) has an entry on "reproduction, reproduire," which defines it as "the action by which a thing is newly produced, or grows a second time."[14] With reference to Trembley's experiments, it is then mentioned that reproduction as the "formation of a new part, absolutely similar to that which had been cut" does not accord with the assumption that all generation depends on an egg.

However, *generation* remains the more complex and comprehensive term in the *Encyclopédie* when it comes to the making of new individuals. The entry "génération" (1757), written by d'Alembert himself, explores the rich semantics of the term. First, generation is explained as "the action of producing what did not exist before," or, more precisely, "the change of one body into another which does not conserve any remains of his previous state."[15] From the very beginning, a clear distinction is drawn between generation and creation. Whereas generation is about the "modification" of already existing bodies, creation means a truly "new production."[16] Moreover generation is associated with "genealogy, or the succession of children and descendants that all come from the same phylum,"[17] and it is termed a "synonym with *people, race, nation*."[18] Most importantly, it is also translated into the semantics of reproduction and linked to the Buffonian concept of the species. In physiology, d'Alembert writes, generation means "the ability to reproduce which is attributed to the organized beings."[19] He continues. "It is thus by the means of generation that the chain of succeeding existences of individuals is formed which constitute the real and non-interrupted existence of different species of beings which do only have a limited duration compared to the stage of organization that gives a determined and distinct form to the individuals of every species."[20] In this definition, partly paraphrasing Buffon, we find many of the elements that are crucial for the formation of the concept of reproduction, namely: the association of reproduction with the species, genealogy ("succession"), and time ("duration").

A brief look at the dictionaries of the period, however, shows that the concept of reproduction did not replace older concepts once and for all. In the *Dictionnaire des sciences médicales* (1812) and the *Nouveau dictionnaire d'histoire naturelle* (1816), for example, both terms overlap. The entries on "generation" in both cases explain the term by using the verb *to reproduce* and the noun *reproduction*. In particular, they emphasize the succession of individuals or the perpetuation of "races and species in the course of centuries"[21] and a cosmology of constant "renewal of the universe."[22] *Reproduction* is said to focus on "the relation between the beings that die and those that are born,"[23] and in some cases it is simply said to be a "synonym" for *generation*.[24] However, it is clearly associated with assumptions about the temporality of supra-individual entities such as "races" or "species." In what follows, I take a closer look at how temporality, genealogy, and the supra-individual level were articulated by different authors and in different constellations between the mid-eighteenth and the early nineteenth century.

REPRODUCTION AS MULTIPLICATION: LA METTRIE, MAUPERTUIS, AND BUFFON

In 1744 when Abraham Trembley published his *Memoirs Concerning the Natural History of a Type of Freshwater Polyp with Arms Shaped Like Horns*, he explained that these animals "could be multiplied by being sectioned" and that he had compared this to their natural way of multiplication.[25] The part in which he deals with the latter is entitled *On the Generation of Polyps*. Only when Trembley describes his experiment of cutting a polyp into two parts— in order to see how these parts each transform into another polyp—he uses the term *reproduction* along with the verb *to replicate*. Whereas the "head of the first part . . . is the original head of the polyp that was divided," he writes, it "scarcely differs from a complete creature."[26] In the back part, however, the "edges of the opening" first "fold inward, and the fold which they form serves to close the aforementioned opening. The anterior end then appears simply swollen and usually remains more or less in the same condition until the reproduction destined to take place there is accomplished."[27] *Reproduction* thus designates an asexual process of regeneration that brings forth a new individual.

Already before the publication of the *Memoir*, Trembley's discovery, reported by Réaumur to the French *Académie des Sciences* and the Court, led to vivid debates among intellectuals because it seemed to support a materialist view on generation in so far as it called into question ideas about an indivisible animal-soul. The mathematician Gabriel Cramer, for example, argued that Trembley's discovery proved that the "soul was material and indistinct substantially from its physical organization."[28] Julien Offray de La Mettrie, in *L'homme Machine* (1748), referred to the polyp for similar reasons. He stressed that the multiplication of the polyp was not consistent with preformationist theory, according to which all individuals, including their "souls," were preformed in God's creation, simply waiting to be "developed" or unfolded.

The fact that the polyps reproduce in eight days "into as many animals as there are cut portions," La Mettrie wrote, "makes me sorry for the system of generation held by the naturalists, or rather, it pleases me very much."[29] He concluded that the discovery of the polyp's ability to reproduce "teaches us never to conclude anything general!"[30] The defenders of preformationism, he maintained, at least had to admit that multiple forms of generation existed.[31]

It is striking that both Trembley and La Mettrie refer to processes of propagation mainly as *multiplication* and that their main argument against preformationism is an argument from multiplication, that is, about acknowledging multiple forms of generation. This also holds for Maupertuis's *Vénus Physique*, published three years before *L'homme Machine* in 1745. The polyp, Maupertuis writes, "in order to multiply itself . . . only needs to be cut in pieces. That part with the head grows a tail while the end with the tail produces a head, and those parts that have neither head nor tail produce both."[32] As with La Mettrie, Maupertuis's argument is about multiple forms of generation and directed against those naturalists who want to reduce all generation to simple *developments* and thus preformation. *New productions* is the term that he opposes to the term *development*, which in the context of preformation theory had no temporal meaning. In particular, Maupertuis bemoaned the understanding of time prevalent in preformationism, which does not account for the succession of time, but instead assumes that God has formed all animals "all at once."[33] To contest this static concept of time, Maupertuis asks if there may be even for God "a difference between the time that we consider to be the same and that which succeeds."[34] Thus, the acknowledgment of the specific form of polyp propagation not only challenges the dominant understanding of generation but also that of time. In particular, it calls into question the idea of time without change, or a mythical time, in which all temporal processes are determined or absorbed by an eternally existing origin. The argument for multiple forms of generation, or "multiplication," thus implies an argument for the multiplicity of time in contrast to the "same time" of creation.

Throughout his discussion of Trembley's discovery, Maupertuis uses the verb *to reproduce*, but the noun *reproduction* only appears in the *Système de la Nature* (1751). However, he links the verb *to reproduce* to several important issues. First, the new understanding of generation he advances takes into account the "equal" contribution of both sexes to the production of a new individual. This idea of equality is, second, a prerequisite for Maupertuis's reflections on hereditary resemblance. "The child is born sometimes with the father's features and sometimes with the mother's. It is born with their faults and their habits and seems even to have inherited their tendencies and mental characteristics."[35] Third, Maupertuis confirms his understanding of generation and hereditary transmission by pointing, among other things, to the "mixture" of people with different skin color.[36] In this way the ideas of succession, descent, and diachronic kinship are crucial when it comes to understanding the process of the emergence of new "species" or "races." Maupertuis argues that "in order to create species from races that become established, it is actually necessary to have the same types unite for several generations. The parts suitable to recreate the original trait, since they are numerous in each generation, are either lost or remain in so small a quantity that a new chance event would be needed to reproduce the original species."[37] Maupertuis, by linking reproduction with issues of gender, hereditary transmissions, monstrous birth, and racial differences, establishes a set of ideas that is crucial for the later debates and politics of reproduction.

Though in Maupertuis, reflections on *reproduction* constitute a new epistemic-political horizon in which the concept is linked to *race*, *gender*, *heredity* as well as to a reconceptualization of time, Buffon is said to have been the first who "gave a wider meaning" to the term *reproduction*.[38] Indeed, in the chapter "De la Reproduction en générale" in volume two of the *Histoire Naturelle* (1749), *reproduction* figures as a comprehensive concept comprising processes of nutrition and generation. Right from the start, the concept is linked to *production*, *succession*, and *species*. It is defined as the "power of producing similarity; the chain of succeeding existences of individuals which constitutes the real existence of the species."[39] Like LaMettrie and Maupertuis, Buffon, by using the term *reproduction*, stresses that there are multiple ways of generating new living beings:

> [O]ne has to mistrust these absolute axioms, these sayings of physics which so many people have wrongly used as principles; for example, there is no fertilization outside the body, *nulla foecondatio extra corpus*, all life comes from an egg, all generation presupposes sexes, etc. One should never take these maxims in an absolute sense.[40]

However, in contrast to Maupertuis, Buffon does not focus on multiplicity when it comes to the temporality of reproduction. Instead he draws attention to processes of renewal and to an eternal logic of destruction and reemergence, or of life and death. In this, Buffon first establishes a genealogical perspective that surpasses the individual. According to him, the cause that reproduces the organization of beings "borrows from destruction itself the means for operating reproduction."[41] Destruction, for Buffon, means to separate the parts or "organic molecules," of which the organism is composed. There is, however, also an "active power" at work that is geared to "reunite" these molecules, which operates in nutrition or in the "assimilation" of matter by plants and animals, as well as in reproduction.[42] Here *reproduction* is therefore not only about the descent of individuals from their parents but also about an eternal succession of destruction and renewal, so that, as a consequence, the idea of a succession of individuals through reproduction clearly acquires a cosmological meaning.[43]

In particular, this results from the complex interplay between time and timelessness in the *Histoire Naturelle*. As Hans-Jörg Rheinberger argues, Buffon conceives of "a time of becoming and passing, in which the individual beings move"[44] and simultaneously of a time of these individuals—their temporal existence. The species itself, in Buffon, is thus understood as existing beyond time as a "genealogical continuum," though it is constituted by finite elements. "The species as unity of reproduction," Rheinberger states, is "on the one hand, defined through its duration and by preserving itself in time as such," and it is "decisively non-historical."[45] On the other hand, the species "is of a temporal nature in a constitutive way, insofar as it realizes itself through the becoming and passing of the individuals, so that the system is based on the timely structure of its elements: The condition for the preservation of the whole is not the preservation of its parts that are precisely transient, but the reproduction of temporally limited units makes it temporally unlimited."[46] This arrangement of temporality obviously goes far beyond Maupertuis's critique of the eternal simultaneity of the time of creation. It combines a cosmological, static time with a dynamic understanding of time that acknowledges the irretrievable passing of time and the emergence of something new.

REPRODUCTION AS REPETITION: BLUMENBACH, HERDER, AND KIELMEYER

As we have seen in the previous paragraph, the concept of reproduction started to circulate in the mid-eighteenth century with strong references to the discoveries of Abraham Trembley. It comes as no surprise, though, that Johann Friedrich Blumenbach begins his essay *Über den Bildungstrieb und das Zeugungsgeschäfte* (On the formative drive and the business of procreation), with a discussion of "experiments of reproduction" he had carried out on polyps.[47] From these experiments he posits the existence of a specific "drive" that prevails in all living beings. The "formative drive," which soon came to play a major role in the debates on a "science of life,"[48] serves for Blumenbach primarily as a term under which he subsumes processes of procreation (*Zeugung*), nutrition (*Ernährung*), and regeneration (*Wiederersetzung*). As he puts it, these processes are "only modifications of one and the same force" that builds, preserves, and repairs.[49] For the most part, he uses the term *reproduction* in a narrow sense, only to designate the regeneration of body parts and not the making of new individuals. Clearly, *generation* is the more comprehensive term that Blumenbach also uses to explain nutrition and reproduction. While nutrition is conceived of as "continued generation," reproduction is a "repeated but only partial generation."[50] So, although reproduction is only a subform of generation for Blumenbach, its specificity lies in the repetitive dimension.

In contrast to Blumenbach's conceptualization of the "formative drive," which is conceived as an equivalent to Newton's gravitational force, Herder's vitalism or *hylozoism* has a pronounced philosophical aspect.[51] However, in his *Ideen zur Philosophie der Geschichte der Menschheit* (Ideas for the philosophy of the history of mankind), Herder also introduces a specific force common to all living beings with reference to Trembley's discoveries. Criticizing the idea of "preformed germs," he postulates—from the polyp's ability to regenerate (*sich erstatten*) and to supplement itself (*sich ergänzen*)—the existence of "organic forces,"[52] "organic vital forces,"[53] or, as he also puts it, "a life sustaining or regenerating organic omnipotence" (das Leben haltende oder erstattende organische Allmacht).[54] Moreover, Herder, in contrast to Blumenbach, elaborates on the concept of reproduction itself. He develops a theory of different forces, including sensibility and irritability, which are arranged differently in different organisms. According to him, the "force of reproduction" (*Reproduktionskraft*) loses itself (*verliert sich*) in the more perfect creatures, whereas sensations and ideas increase until, in the human, they develop into reason. In the lower creatures, however, the forces of generation (*Fortpflanzung*) and regeneration (*Wiedererstattung*) are powerful. Each organism is thus understood as a relation of forces and the hierarchical order of living beings as structured through the differences between these specific arrangements of forces. This, for Herder, is linked to the idea of an analogous development of individual beings and of the order of living beings in general—an idea that has become prominent under the title of *recapitulation*. Accordingly, every human being undergoes all states that appertain to less complex forms of living beings.

As Peter Hanns Reill has argued, Herder thereby combines two models of change: a linear one that assumes a progression from lower to higher creatures by an "intensification of complexity" and an analogical model.[55] According to the latter, Herder argues "that living creatures evince 'a certain unity of structure,' explicable only by the existence of what he called the *Haupttypus* (principal type), and *Hauptform* (principal form or prototype)."[56] Both models of change operate on a supra-individual level and situate the reproduction of individuals in a larger genealogical framework where the order of nature is repeated and mirrored by the development of the individual. Thus, *reproduction* is not only at the heart of Herder's idea of vital forces and of the differences and hierarchies in the world of living beings, it articulates a cosmological theory which accounts for ideas of the correspondence of the micro- and macro-cosmos[57] and of a circular temporal structure of nature. These ideas were to circulate in the following debates.

One of the authors who took up Herder's central ideas was Carl Friedrich Kielmeyer. In 1793, Kielmeyer gave a lecture at the *Hohe Karlsschule* in Stuttgart where he had studied and taught. In his *Karlsschulrede*, which was widely read in the years that followed, Kielmeyer developed a compensation theory of forces and a theory of recapitulation similar to Herder, though he did not mention him. According to Kielmeyer, each organism is defined by a specific relation of the forces of irritability, sensibility, secretion, propulsion (i.e., the ability to move and to distribute fluids "in the firm parts according to a certain order"),[58] and reproduction. Although it is only one force among others, the "force of reproduction" plays a major role in Kielmeyer's theory. Ultimately, his compensation model of forces and the idea of recapitulation, that is, the assumption that "the human and the bird, too, are plant-like at their first stage,"[59] are based on observations of processes of reproduction.

Kielmeyer starts from the assumption of a "multiformity of appearances" of this force, which asserts itself in a range of temporalities.[60] According to him, the "force of reproduction" can be "eternally uniform" or "changeable"; it can "persist unfrayed for centuries" or be determined in its changes by a "temporal inflection."[61] From this multiformity, he derives three laws of "compensation,"[62] and, in a further step of abstraction, the "general law" of compensation.[63] According to the latter, the forces of sensibility and irritability decrease when the force of reproduction increases and vice versa. As in Herder, the compensation theory is linked to the idea of recapitulation. The laws of reproduction thus "coincide with the force through which the series of different organizations of the earth has been called into being."[64] As a consequence—and as Kielmeyer had pointed out already at the beginning of his speech—the life of the individual is integrated within a "larger system,"[65] the "life of the species," and this, in turn, is part of the "life of the organic world."[66]

Clearly, Kielmeyer's understanding of reproduction has some cosmological implications. He is not only concerned with the attempt to find "a material cause for the explanation of the phenomena of development," be it with respect to the individual or to the "generation of the organizations of our earth."[67] As for Herder, the idea of reproduction as recapitulation implies assumptions about the belonging of each individual to an unmanageable ancestry and, in a synchronic perspective, to a hierarchically structured order of living beings. This articulation of reproduction rivals the narration of creation insofar as Kielmeyer equates the "force of reproduction" with the primordial force that "called into being" all living beings on earth.[68] In a letter to Carl Friedrich Eschenmayer from 1799, in which he contemplates Eschenmayer's book *Sätze aus der Naturmetaphysik auf chemische und medizinische Gegenstände angewandt* (Sentences from the metaphysics of nature, applied to chemical and medical issues), Kielmeyer also admits that he had often had the "idea of a cosmogony in analogy to a theory of the formation of an origin."[69]

However, the cosmological dimension in Kielmeyer remains epistemologically restricted. This is particularly clear in his dealing with issues of time. Although the organic world in general is structured by a circular form of time that is derived from and geared toward a mythical origin, this is countered by Kielmeyer's interest in the multiplicity of forms of reproduction and in multiple forms of temporality and change. This interest, which contradicts any homogenization of time, also appears in Kielmeyer's *Entwurf zu einer vergleichenden Zoologie* (Draft for a comparative zoology, 1814). There Kielmeyer highlights his interest in synchronic differences as well as differences between organisms "at different times," and he carefully distinguishes between different forms of temporal change, for example, between changes that are repeated and those that are not repeated in an animal body.[70] Kielmeyer, as Peter Hanns Reill has argued, "believed no simple uniform developmental pattern accounted for species change and permanence."[71] Rather, he was inclined to complexity and multiplicity.

REPRODUCTION AS RETURN: SCHELLING, GÖRRES, HEGEL

In the *Naturphilosophie* of Schelling, Görres, and Hegel things look different. *Reproduction* here is introduced along with an ontological dualism of the sexes that structures the whole of organic nature—if not the universe—and also lies at the heart of philosophical narrations of a return to the origin. However, *Naturphilosophie*, which from the 1830s on was increasingly criticized for its speculative design, had close connections to the emerging life sciences in the decades around 1800. Far from being opposed to science, it was "perceived as offering natural history a rise in status from a mere appendage of the medical faculty to full membership alongside mathematics, philology, and physics in a higher philosophical faculty."[72] The circulation of the concept of reproduction is a good example of the ambivalent status of scientific knowledge in philosophical discourse, which also rose in status by embracing scientific knowledge.

When Schelling introduced the project of *Naturphilosophie* in his writings between 1797 and 1799, he derived his leading question from his and Fichte's critique of Kant. The main argument they had put forward was that, in Kant's philosophy, the status of self-consciousness—which was said to be the "highest principle"—nonetheless remained unclear. It was Fichte's and Schelling's philosophical project to elaborate a comprehensive understanding of subjectivity that allowed it to be the connecting principle for all branches of Kant's philosophy and thus the foundational principle of philosophy in general. Schelling, however, then rearticulated this task of "transcendental philosophy" as a "natural philosophy." From its inception, *Naturphilosophie* was therefore mainly a philosophical enterprise seeking to explain the "principle of being," which, according to Schelling, reveals itself in every natural object. Building on Fichte's theory of the "I" as absolute activity, which posits itself and the world, Schelling understood being in a strictly dynamic and active sense. He conceived of it as the "highest constructing activity, which, although never itself an object, is the principle of everything objective."[73] Thus starting from an understanding of nature as infinite activity or productivity, the main problem for a philosophy of nature then was "not to explain the active in Nature . . . but the *resting, permanent.*"[74]

In Schelling's attempt to explain the existence of natural entities, which come into being by way of inhibiting the productive, infinite process of nature, reproduction plays a central role. First, Schelling assumes, in contrast to Blumenbach, that the formative drive unfolds in two directions: female and male. As a consequence, sexual difference is assumed to structure the whole of the organic world and finally nature in general. Sexual difference, he claims, must "be taken as a point of departure everywhere in organic nature,"[75] which, however, is geared toward unification and the generation of a common product: a new individual. Although the resulting individual appears to be fixed as an entity in itself, it is not. In the "endless circulation," in which nature "returns into itself," the individual vanishes for the sake of the species.[76] *Reproduction*, thus, not only explains the production of an individual through sexual propagation, but more significantly, it is the structuring principle of the process of nature in general. This means that Schelling does not only refer to the "force of reproduction" to understand the organism according to a model of forces based on insights by Herder, Blumenbach, and Kielmeyer. Reproduction first and foremost articulates his philosophical narrative and the nostalgia for the return to the origin.

This means that not only individuals and species reproduce themselves "to infinity."[77] Insofar as nature is conceived of as "one organism,"[78] nature as a whole reproduces "itself," and the integrated process of production, destruction, and reproduction is seen as the fundamental dynamics of the universe. "The force which appears as active in reproduction is a force infinite in its nature, for it is joined to the eternal order of the universe itself."[79] This order is a "*continuous return of Nature into itself.*"[80] *Nature* here stands for a primordial genealogical power that constantly brings forth and destroys individuals. All living beings are in every case already subsumed by this power and thus to the collective entities of the sex and the species. Individuals, accordingly, are nothing else than "misbegotten attempts" to represent the Absolute,[81] and nature "contests the individual,"[82] which is thus always "submitted to a higher order."[83] The synchronic and diachronic bonds among living beings are all constituted by this circular process—by *Nature* in general. The predominance of a supra-individual perspective on the production and generation of new individuals could not be clearer. Building on the ideas of Buffon, Herder, and Kielmeyer, but overcoming their ambivalences, *reproduction* in Schelling is transformed into the main principle of a cosmic dynamic. Accordingly, and in sharp contrast to the language of activity and dynamics, "time" in Schelling appears as something static. Ultimately, all processes are geared toward the "highest point" where "no time elapses."[84]

Many of these features are repeated again and again in post-Schellingian *Naturphilosophie,* in particular, the "obsession with the idea of return" and the "de-temporalization" of time.[85] Joseph Görres, for example, who among other things tried to devise a comparative physiology as outlined by Schelling, reappropriated important features of his philosophy of nature. Significantly, this includes the idea of a fundamental duality of the sexes that structures the universe. Instead of stressing an understanding of reproduction as a process of return that relies on the contribution of both sexes, Görres elaborates on the difference between *productive* and *eductive* activity. According to his *Aphorismen über die Kunst* (Aphorisms on art, 1802), nature generates a diremption of itself into "an absolute productive and absolute eductive activity."[86] This duality is heavily invested with an Aristotelian understanding of generation that Görres transposes into the language of *Naturphilosophie*. The hierarchical difference between the sexes, seen in terms of the matter/form distinction, is reaffirmed in light of the ideal of gender complementarity and reunification. Similar to Aristotle, the female contributes "plastic (*bildsame*) organic matter," the male "gives form to the formless," so that in the end both "stream (*fliessen*) into one another for the new life."[87]

Although Görres uses the term "reproduction" only on a few occasions, it is clear that he adopts the Schellingian model of an eternal, sexualized dynamic. The sex, he writes "reproduces itself in the antagonism of masculinity and femininity, and the species protects itself against expiration."[88] Like Schelling, he takes sexual propagation as a model for a larger circular process: the process of "life." "The product of each single act of procreation (*Zeugung*) is only one factor of the next act of procreation (*Zeugungsakt*)," so that the "life wheel writhes (*wälzt sich*) through time."[89] Sexual propagation again is at the heart of a cosmological narration of an eternal attempt to return to the origin.

Hegel's philosophy of nature, finally, shares important characteristics with Schelling's and is also conceived as a narration of return. Hegel starts from the assumption of a differentiation between nature and spirit and declares it to be "the specific character and the goal of the Philosophy of Nature that Spirit finds in Nature its own essence, i.e. the Notion, finds its counterpart in her."[90] The whole enterprise has a circular structure, since it is understood as a "path of return," in which the separation of nature and spirit is overcome, and the spirit, or God (the "all-sufficient") takes "back into itself" what it had first produced as its other.[91] Within this narration, Hegel establishes a dynamic model of nature, so that the "stages" of minerals, plants, and animals arise from one another, each "being the proximate truth of the stage from which it results."[92]

Reproduction here plays a central role in that it marks the transition from the realm of minerals to the realm of plants. With plant nature, Hegel writes, "the truly organic form of life" begins, as "the vital principle (*Lebendigkeit*) gathers itself into a point and this point sustains and produces itself, repels itself, and produces new points."[93] However, the process of reproduction in plant nature still remains restricted insofar as it does not rely on a difference of the sexes. "Reproduction," Hegel argues, is still "not mediated by opposition,"[94] which only appears in animal nature. Or, in other words, only the animal organism is essentially reproductive—"this is its actuality."[95] Like

Schelling, Hegel is inspired by Kielmeyer's compensation model of forces. Sensibility and irritability thus characterize the higher forms of life while "the lower forms of life remain reproduction."[96] "Thus there are animals," he continues, "which are nothing more than a reproductive process—an amorphous jelly, an active slime which is reflected into itself and in which sensibility and irritability are not yet separated."[97] The contrast to Oken's ideal of pure masculinity, where the power of reproduction is claimed to be the "male principle," or "the polyp," could not be more obvious.[98] Modes of reproduction that lack sexual difference are at the lowest stage—nothing other than "active slime." Real reproduction for Hegel, as for Schelling, presupposes two sexes, and although their arrangement is hierarchical, there is a sense of complementarity and mutual desire.

However, the relation of the sexes from the outset is determined by the exigencies of the species. Like Schelling, Hegel conceives of reproduction as structuring the "genus-process" (*Gattungsprozess*), and here again the individual is clearly in a subordinate position. The species for Hegel is "the negative power over the individual . . . realized through the sacrifice of this individual which it replaces by another."[99] However, the "extinction of the singularity"[100] in death is not only crucial for the eternal perpetuation of the species, but at the same time, it indicates the transition from organic nature to spirit. Therefore, organic nature, as Hegel puts it, "ends, for with the death of the individual the genus comes to its own self and thus becomes its own object: this is the procession of spirit."[101] In this process, the return to the origin shall succeed. *Nature* recognizes itself as idea and "reconciles itself with itself."[102] Reproduction, however, has become a process that organizes this process of return. Although it only plays a central role on the level of organic nature, which is to be overcome, it is, as with Schelling and Görres, part of a cosmological narration that transposes sexual difference into a dualism structuring the universe.

CONCLUSION

I started with the assumption that the emergence and circulation of the concept of reproduction, from the mid-eighteenth century on, indicates a cultural concern with genealogy that is part of a broader transformation of experiences of time and of belonging to natural and social orders. The material discussed earlier shows that the concept is, in various ways, invested with issues of genealogical belonging. In La Mettrie, Maupertuis, and Buffon, *reproduction* is articulated with multiplicity, both in the sense that reproduction is understood *as* multiplication and also in the sense that the concept of reproduction is used to challenge predominant assumptions about generation and time by way of ideas of multiplication. With Buffon, however, a cosmological dimension is attached to the concept of reproduction, and the individual—insofar as it belongs to the species—is situated in an eternal process of becoming and passing.

Conceptualizing the individual with relation to the species and within the encompassing process of the organic world and nature in general becomes an important theme for Herder and Kielmeyer, although their texts also display a concern with multiplicity and, in particular, with the coexistence of different forms of time and change. Cosmological ideas of an eternal circuit constituting a kind of universal kinship of all living beings are epistemologically restricted in that they are contrasted with other forms of temporality. Only in the interventions of *Naturphilosophie* this cosmological dimension becomes central. In particular, sexual propagation becomes the central model on which their narrations of a return to an imaginary origin build. Sexual difference, now articulated in terms of dualism and antagonism, is treated as a fundamental principle of the universe. As a consequence, there is no individual existence that can escape this dualism because the individual is always already overpowered by the exigencies of the species and of *Nature*.

The move toward cosmology and gender dualism thus goes hand in hand with the constitution of a species-oriented perspective. This increasing interest in the regularity of long-term processes and continuity, however, did not only constitute a new scientific perspective. The emergence of such a species-oriented perspective also contributed to the constitution of a biopolitical gaze. Increasingly, individuals were perceived as belonging to a species, a sex, or a race in a biological way that subjugated them to these larger entities. Although the articulations of the concept of reproduction that I have studied in this chapter do not explicitly refer to strategies of population politics that were developed in the period, the scientific and philosophical articulations of reproduction resonated with the social and political transformations of the period. In particular, the emergence of and understanding of human beings as subjects of biological processes fostered a new "naturalized" understanding of kinship, genealogy, and belonging that came to play a crucial role in the restructuring of social relations.

NOTES

1. François Jacob, *The Logic of Life: A History of Heredity* (London: Penguin, 1989), 19–20.
2. Staffan Müller-Wille, Hans-Jörg Rheinberger, "Heredity—the Formation of an Epistemic Space," in *Heredity Produced: At the Crossroads of Biology, Politics, and Culture, 1500–1870*, ed. Staffan Müller-Wille and Hans-Jörg Rheinberger (Cambridge, MA: MIT Press, 2007), 6.
3. Reinhard Koselleck, "The Eighteenth Century as the Beginning of Modernity," in *The Practice of Conceptual History: Timing History, Spacing Concepts* (Stanford: Stanford University Press, 2002), 165.
4. Koselleck, "The Eighteenth Century."
5. In the German context, the term *Geschlecht* itself underwent a change. Until the late eighteenth century it did not designate first and foremost "sex" and "gender" but rather a larger kinship entity—a line of ancestors. The *Zedler'sche Universallexikon* thus presents "Genus, Famille, Maison" as equivalents, and explains it as "parentage (*Abkunft*), descent (*Abstammung*) and lineage (*Herkommen*) of one human form the other." See *Zedler'sches Universallexikon*, http://www.zedler-lexikon.de/, accessed March 23, 2011.
6. Ludmilla Jordanova, *Nature Displayed: Gender, Science, and Medicine, 1760–1820* (London and New York: Longman, 1999), 67.
7. Jordanova, *Nature Displayed*, 63.
8. Ibid., 1.
9. Ohad Parnes, Ulrike Vedder, and Stefan Willer, *Konzept der Generation: Eine Wissenschafts- und Kulturgeschichte* (Frankfurt/Main: Suhrkamp, 2008), 29. All translations in this chapter, unless stated otherwise, are my own.
10. Friedrich Kaulbach, "Produktion, Produktivität," in *Historisches Wörterbuch der Philosophie*, ed. Karlfried Gründer and Henning Ritter (Darmstadt: Wissenschaftliche Buchgesellschaft, 1976), 1418–1426.
11. Johann Micraelius, *Lexikon Philosophicum Terminorum Philosophis* (Stettin: Mamphrasius, 1661), http://www.uni-mannheim.de/mateo/camenaref/micraelius.html, accessed October 20, 2012, p. 1139.
12. *Encyclopédie ou dictionnaire raisonné des sciences, des arts et des métiers*, s.v. "production," http://encyclopedie.uchicago.edu/, accessed August 15, 2011.
13. For the changing meanings of the terms *Zeugung* and *Fortpflanzung* around 1800 see the chapter of Jocelyn Holland in this volume. As

Holland shows, *Fortpflanzung* became associated with succession and duration and thus acquired a meaning that is very close to that of *Reproduktion*.
14. *Encyclopédie*, s.v. "réproduction, réproduire."
15. *Encyclopédie*, s.v. "génération."
16. Ibid.
17. Ibid.
18. *Encyclopédie*, s.v. "génération." For the double meaning of *generation* which designates first a temporally defined cohort of individuals and second the process of bringing forth new individuals see Parnes, Vedder, Willer, *Konzept der Generation*.
19. *Encyclopédie*, s.v. "génération."
20. Ibid.
21. *Dictionnaire des sciences médicales*, s.v. "génération," http://www2.biusante.parisdescartes.fr/livanc/?cote=47661x01&do=chapitre, accessed August 15, 2011.
22. *Nouveau dictionnaire d'histoire naturelle, appliquée aux arts, à l'agriculture, à l'économie rurale et domestique, à la médecine, etc.*, s.v. "génération," http://www.archive.org/details/nouveaudictionna02metc., accessed August 15, 2011.
23. *Nouveau dictionnaire d'histoire naturelle*, s.v. "reproduction."
24. Ibid.
25. Abraham Trembley, "Memoirs Concerning the Natural History of a Type of Freshwater Polyp with Arms Shaped Like Horns" (1744), in *Hydra and the Birth of Experimental Biology*, ed. and trans. Silvia G. Lenhoff and Howard M. Lenhoff (Pacific Grove: Boxwood Press, 1986), 93.
26. Abraham Trembley, "Memoirs," 142.
27. Trembley uses the French word *reproduction*. Lenhoff and Lenhoff, however, translate this as *régénération*. Abraham Trembley, "Memoirs," 143.
28. Aram Vartanian, "Trembley's Polyp and La Mettrie," in *Journal for the History of Ideas* 11, 3 (1959):259–286, 264.
29. "Les Polypes . . . se reproduisent dans huit jours en autant d'Animaux, qu'il ya de parties coupées. J'en suis fâché pour le système des Naturalistes sur la génération, ou plutôt j'en suis bien aisé." Julien Offray de LaMettrie, *L'Homme Machine* (Utrecht: Presses de l' Imprimerie Bosch, 1966), 124. I follow the translation of Vartanian in "Trembley's Polyp," 271.
30. "car cette découverte nous apprend bien à ne jamais rien conclure de général," LaMettrie, *L'Homme Machine*, 124. I follow the translation of Vartanian in "Trembley's Polyp," 271.

31. Indeed, Bonnet, one of the main advocates of preformation theory, "was very well aware of the several problems that the polyp's hydra-like regeneration created for the defenders of the animal-soul but refrained at the time from attempting a solution" (Vartanian, "Trembley's Polyp," 265). For the different forms of preformationist theory, see Roselyne Rey, "Génération et hérédité au 18e siècle," in *L'Ordre des Caractères: Aspects de l'hérédité dans l'histoire des sciences de l'homme*, ed. Jean-Louis Fischer and Claude Bénichou (Paris: Sciences en Situation, 1989), 7–41.
32. Pierre-Louis Moreau de Maupertuis, *The Earthly Venus*, trans. Simone Brangier Boas (New York and London: Johnson Reprint Corporation, 1966), 39.
33. Maupertuis, *The Earthly Venus*, 42.
34. The French original reads: "y-a-t'il même pour Dieu quelque différence entre le temps que nous regardons comme le même, & celui qui se succède?" Brangier Boas translates: "For God, is there any real difference between one moment in time and the next?" Maupertuis, *The Earthly Venus*, 42.
35. Ibid., 43.
36. On Maupertuis' discussion of the White Negro, the phenomenon of albinism in Africans, see the chapter by Staffan Müller-Wille in this volume.
37. Maupertuis, *The Earthly Venus*, 80.
38. Jacob, *The Logic of Life*, 72.
39. Georges-Louis Leclerc, Comte de Buffon, *Histoire Naturelle*, ed. Pietro Corsi and Thierry Hoquet, http://www.buffon.cnrs.fr/, accessed October 20, 2012, vol. 2, 18.
40. Buffon, *Histoire Naturelle*, vol. 2, 33.
41. Ibid, 41.
42. Ibid.
43. On the relation of Buffon's cosmological theory with his concept of racial degeneration see Philip Sloan, "The Idea of Racial Degeneracy in Buffon's 'Histoire Naturelle,'" in *Racism in the Eighteenth Century*, ed. Harold Pagliaro (Cleveland: Case Western Reserve University, 1973), 293–321. On the role of "degeneration" in Gottfried Reinhold Treviranus' work *Biology*, see the chapter of Joan Steigerwald in this volume.
44. Hans-Jörg Rheinberger, "Buffon: Zeit, Veränderung und Geschichte," in *History and Philosophy of the Life Sciences* 12 (1990):206.
45. Rheinberger, "Buffon," 206.
46. Ibid., 207.

47. Johann Friedrich Blumenbach, *Über den Bildungstrieb und das Zeugungsgeschäfte* (Stuttgart: G. Fischer, 1971), 9.
48. In Kant's *Critique of Judgment*, for example, the *Bildungstrieb* "serves as a model . . . for the way mechanical and teleological explanation can be united." Robert Bernasconi, "Kant and Blumenbach's Polyps: A Neglected Chapter in the History of the Concept of Race," in *The German Invention of Race*, ed. Sara Eigen (Figal) and Mark Larrimore (Albany: State University of New York Press, 2006), 78.
49. Bernasconi, "Kant and Blumenbach's Polyps," 19.
50. Ibid.
51. As John Zammito shows, the debates about "*hylozoism*, the need to see matter and spirit in continuity . . . carried Herder decisively to Spinoza" who, via Herder, became one of the most important authors in German idealism. John H. Zammito, *Kant, Herder, and the Birth of Anthropology* (Chicago and London: University of Chicago Press, 2002), 317.
52. Johann Gottfried Herder, *Ideen zur Philosophie der Geschichte der Menschheit* in *Werke*, ed. Martin Bollacher (Frankfurt am Main: Deutscher Klassiker-Verlag, 1989), vol. 6:90.
53. Herder, *Ideen zur Philosophie*, 92.
54. Ibid.
55. Reill, *Vitalizing Nature*, 190.
56. Ibid., 189.
57. The human, according to Herder, contains all forms of existence that preceded him: "His blood and his many named components are a compendium of the world: calcium and earth, salts and acids, oil and water, forces and vegetation, irritability and sensation are originally united and intermingled with each other in him." Herder, *Ideen zur Philosophie*, 165. I follow the translation of Reill in *Vitalizing Nature*, 188.
58. Carl Friedrich Kielmeyer, *Ueber die Verhältniße der organischen Kräfte unter einander in der Reihe der verschiedenen Organisationen, die Gesetze und Folgen dieser Verhältniße* (Marburg an der Lahn: Basilisken Presse, 1993), 10.
59. Kielmeyer, *Ueber die Verhältniße*, 36.
60. Ibid., 26.
61. Ibid., 25.
62. His first law states that the number of individuals formed decreases, commensurate with the "size of the generating, or . . . of the generated individuals as they appear after birth" (Kielmeyer, *Ueber die Verhältniße*, 28).

According to the second law, an increase in the number of new individuals correlates to a decrease in the complexity of their physical structure and to a shorter period necessary for their formation (Kielmeyer, *Ueber die Verhältniße*, 30). And the third law posits that "the less the force of reproduction manifests itself in many new individuals, the more it manifests itself in transformations that the body survives, or in unusual artificial reproduction, or both at the same time, or in an undefined growth, or in greater deviation in the newly generated formations" (Kielmeyer, *Ueber die Verhältniße*, 33).

63. Ibid., 35.
64. Ibid., 38.
65. Ibid., 4.
66. Ibid., 5.
67. Ibid., 39.
68. Ibid.
69. Carl Friedrich Kielmeyer, "Über mechanische, chemische und organische Bewegung" in *Natur und Kraft* (Berlin: W. Keiper, 1938), 58.
70. Carl Friedrich Kielmeyer, "Entwurf zu einer vergleichenden Zoologie," in *Natur und Kraft* (Berlin: W. Keiper, 1938), 21.
71. Reill, *Vitalizing Nature*, 195.
72. Nicholas Jardine, "*Naturphilosophie* and the Kingdoms of Nature," in *Cultures of Natural History*, ed. Nicholas Jardine, Jim A. Secord, and Emma Spary (Cambridge: Cambridge University Press, 1996), 243.
73. Friedrich Wilhelm Joseph Schelling, *First Outline of a System of the Philosophy of Nature*, trans. Keith R. Peterson (Albany: State University of New York Press, 2004), 14.
74. Schelling, *First Outline*, 17. For a detailed account of Schelling's project of *Naturphilosophie* and the constructions of sexual difference in Schelling and Hegel see the chapter of Alison Stone in this volume.
75. Schelling, *First Outline*, 36.
76. Ibid., 42.
77. Ibid., 46.
78. Ibid., 54.
79. Ibid., 130.
80. Ibid., 93.
81. Ibid., 35.
82. Ibid.
83. Ibid., 139.
84. Ibid., 40.

85. Reill, *Vitalizing Nature*, 216.
86. Joseph Görres, "Aphorismen über die Kunst," in *Naturwissenschaftliche, kunst- und naturphilosophische Schriften I, 1800–1803* (Köln: Im Bilde-Verlag, 1932), 65. *Eductio* is a scholastic term that means the generation of "forms" from the potentialities of "matter."
87. Görres, "Aphorismen," 109. On the extremely explicit gender hierarchies in post-Schellingian *Naturphilosophie* see the accounts of Lorenz Oken and Carl Gustav Carus by Peter Hanns Reill and Florence Vienne in this volume.
88. Görres, "Aphorismen," 140.
89. Ibid., 110.
90. Georg Friedrich Wilhelm Hegel, *Philosophy of Nature, Being Part Two of the Encyclopedia of the Philosophical Sciences*, trans. A. V. Miller (Oxford: Clarendon Press, 1970), §246, p. 13.
91. Hegel, *Philosophy of Nature*, §247, p. 14.
92. Ibid., §249, p. 20.
93. Ibid., §342, p. 303.
94. Ibid., §345, p. 312.
95. Ibid., §353, p. 358.
96. Ibid.
97. Ibid.
98. Oken's ideal is asexual reproduction and his nostalgia for the origin is a nostalgia for a state of pure masculinity. See Peter Hanns Reill and Florence Vienne in this volume.
99. Hegel, *Philosophy of Nature*, §348, p. 346.
100. Ibid., §367, p. 411.
101. Ibid.
102. Ibid., §376, p. 444.

2

ORGANIC MOLECULES, PARASITES, *URTHIERE*

The Controversial Nature of Spermatic Animals, 1749–1841

FLORENCE VIENNE
TRANSLATED BY KATE STURGE

The closing years of the eighteenth century have often been described as a phase of transition from a descriptive and classifying "natural history" to a "science of life" that sought general laws of organic development.[1] This chapter approaches that period of transformation through contemporary investigations of the microscopic animals of semen. In 1677, Antoni van Leeuwenhoek (1632–1723) became one of the first to observe and describe these animalcules. His discoveries led him to propose that sperm contained the preformed embryo.[2] Leeuwenhoek's hypothesis found several adherents, but by the second half of the eighteenth century no naturalist continued to defend the view that the germs, which God had put into the world at the moment of Creation, were preexistent—"encased" within the spermatic animalcule.[3] However, if the animalcules were not the preexisting germs or embryos of future organisms, what was the *raison d'être* of these peculiar inhabitants of the semen? Were they even animals or living beings at all? From the mid-eighteenth century on, such questions occupied not only the proponents of preexistence theory, but also its adversaries. These opponents are the focus of this chapter. More precisely, I discuss the theories of generation formulated by the naturalist Georges-Louis Leclerc, Comte de Buffon (1708–1788), the physiologist Johann Friedrich Blumenbach (1752–1840), and the physician and Naturphilosoph Lorenz Oken (1779–1851). All three were seeking to explain the generation of new life in ways that no longer rested on God's act of creation, but their answers led them to fundamentally divergent views of the nature and function of spermatic animals. Buffon regarded these as initial combinations of "organic molecules"; Blumenbach as parasites of the semen; and Oken as primordial animals or *Urthiere*. By comparing these positions, I trace the emergence of different visions of organic nature that were associated with different concepts of gender relations. In addition, I discuss exemplary cases from the development of cell biology that demonstrate to what extent these conflicting views continued to inform the science of life through the nineteenth and well into the twentieth century.

THINKING THE ORGANIC WORLD AS A UNITY: BUFFON AND BLUMENBACH

The second volume of Buffon's natural history, *Histoire naturelle, générale et particulière*, first published in 1749, is dedicated to the issues of reproduction and generation. It includes a report on extensive microscopic investigations of the seminal fluid of human cadavers and of various animals, both male and female. Buffon explains that he only undertook these studies after having formulated his theory of generation,[4] a crucial aspect of which was the notion that nature consisted of an infinite number of organic "particles" or "molecules." When acted on by a force—which Buffon likened to gravity—these "organic particles" united to form both elementary and more complex organized bodies.[5] It was precisely in the hope of "recognizing," as he puts it, these "living organic particles" that he took up his investigations of seminal fluid.[6] Little wonder, then, that what Buffon "saw" when looking at the semen through the microscope deviated critically from the engravings prepared by Leeuwenhoek. For example, he notes that Leeuwenhoek's depictions of the so-called spermatic animals generally made them too thick and too long.[7] Neither did they move on their own momentum, as Leeuwenhoek had assumed—rather, this movement was a product of the experimental practices (such as the hand trembling) and the liquids in which they were suspended.[8] Above all, however, Buffon was interested in the animalcules' thread-like appendages, which Leeuwenhoek (erroneously in Buffon's view) had described as "tails." For Buffon, this "tail" could not be the body part of an animal. When he placed the seminal fluid of human cadavers under the microscope one to fourteen hours after extraction, he saw that the thread became shorter and shorter. It increasingly separated from the small moving body and finally disappeared entirely.[9] Buffon observed that the moving bodies' bulk diminished between the tenth and eleventh hour outside the organism. They became little globules, which joined with one another to form a net resembling the "web of a spider besprinkled with drops of dew," with the dew as an enormous number of small globules.[10] Buffon repeated his observations with the seminal fluid of a living dog, "emitted in the natural manner," and with the semen of dissected dogs, rabbits, and rams, taken from the testicles. In all cases he found small moving bodies similar in form and size to those in human seminal fluid. The processes that occurred when Buffon examined the semen microscopically at different intervals were also the same: the detachment of the so-called tail, its transformation into small globules and finally the associations into which those globules entered.[11] What the naturalists had hitherto considered to be animals and living beings, therefore, Buffon reinterpreted as "the first union or assemblages" of "organic particles."[12]

Buffon's proof that the semen did not contain living animals but living organic molecules was of prime importance in confirming his theory of generation. At stake was not only a critique of the animalculist version of preexistence theory. Perhaps to an even greater extent, the objective of his microscopic studies was to refute the theory's ovist version. For this reason, his discussion of reproduction challenged not only the work of Leeuwenhoek and his disciples, but also that of William Harvey (1578–1657) and of his student Reinier de Graaf (1641–1673). Both had emphasized the importance of the ovum in generation.[13] Buffon's aim was to refute Harvey's view that human beings and animals came from eggs; the egg was, he argued, neither the place where the chick was formed, nor did it function as an agent of generation. It could not be understood as an active and essential unit of generation but served solely to nourish the embryo.[14] Indeed, for Buffon the vesicle that de Graaf had described as an *ovum*—due to its resemblance to the *ova* found in the ovaries of hens—was not an egg at all. It was wrong to assume that mammals possessed eggs, and this error had, according to Buffon, led Harvey and de Graaf to the incorrect view that only one sex played a role in generation.[15] He cast doubt not only on this view but, more generally, on all those theories of generation that assumed an asymmetrical contribution to generation by the two sexes. Such theories included the Aristotelian dichotomy between form and matter, according to which the man represented the sole generative force, or efficient principle, whereas the woman supplied matter in the shape of menstrual blood. Buffon gave little credence to Aristotle's notion that seminal fluid contributed no matter but worked like a sculptor forming a piece of marble, and he equally disregarded all other attempts to locate the origin of life exclusively in one of the two sexes.[16] In his opinion, the existence of hybrids (for example, the offspring of a donkey and a horse), and especially the fact that children resemble both their mother and their father, put beyond doubt the fact that "both parents have contributed to the formation of the child."[17]

In search of a different theory based on an equivalence of both sexes' contribution to generation, Buffon postulated the existence of a generative material, common to both the male and the female organism. He adopted the view of Hippocrates and Galen that the female sex was endowed with a seminal fluid analogous to that of the male.[18] Taking his cue from Pierre-Louis Moreau de Maupertuis (1698–1759), he assumed that the two seminal fluids had to mix for generation to take place.[19] Once he had carried out his microscopic studies of semen, Buffon needed to extend his investigations to female animals to confirm his theory of generation. In line with his belief that females possessed not ovaries but testes, he dissected a female dog, found the organ in question—with ease, as he stresses in his description of the experiment—and removed a fluid from it. Examining this fluid under the microscope, he "had the satisfaction of perceiving, at the first glance," that it, too, contained small moving bodies. These were "exactly similar" to the bodies he had found in the seminal fluid of the male dog.[20] Buffon repeated his observations with the seminal fluids of other female dogs; he even mixed them with the semen of a male dog. The moving bodies he had observed under the microscope were so similar as to be indistinguishable.[21] From freshly slaughtered cows, as well, it was possible without difficulty to extract the seminal fluid from the testes, and in this fluid Buffon again and again found active, mobile bodies.[22] He had thus established that the female semen contained the same moving, living organic particles.[23] The two seminal fluids—and this was Buffon's main finding—represented two "equally active" materials.[24]

However, if the generative matter of one sex possessed all the necessary preconditions for reproduction, the objection inevitably arose that the other sex was superfluous. To counter that objection, Buffon offered a series of explanations aiming to demonstrate the necessity of both sexes for generation in humans and animals. The molecules contained in the generative matter of the two sexes came from different parts of the parental organisms and were miniature images of individual parts of the body. It was only through the "assemblage" of the organic molecules of the father and the mother that complete organs and full embryos could be formed—some of them female and others male.[25] In another passage, Buffon proposed that the molecules of one individual could only unfold their full activity through the force or resistance of the molecules of the other. I do not pursue Buffon's thinking on the interplay of the organic molecules of the two sexes any further here.[26] His assumptions regarding the complementary roles of the sexes in reproduction did not lead him to postulate a fundamental distinction or hierarchy between them. On the contrary, Buffon's research aimed to rigorously eliminate all differences between the reproductive bodies and materials. Comparing this with later theories of generation, it is striking that Buffon extends his principle of symmetry to the constitutive elements of both generative materials. The organic molecules of women's and men's seminal fluid were not all identical, because they represented different parts of the body, but they were mutually dependent in their interaction, and in their primary function as life-constituting units they were equally efficient. For Buffon, there were no female and male organic molecules, just as there were no sex-specific generative materials.

In fact, more generally, the presence of the seminal fluid—its production and its function—were not bound up with the body's sexual organs, whether that body was male or female. An important aspect of Buffon's theory of generation was that he saw the seminal fluid as being produced by an excess of nutriment (because women were smaller and took less food, their seminal fluid was present in smaller quantities).[27] Buffon also emphasized that the organic molecules he observed and described in semen did not differ significantly in their strength and effect from those that were present in other plant and animal substances.[28] The difference was merely that semen contained them in more abundance.[29] The organic particles that Buffon identified through his microscopic studies of seminal fluid were, then, not those of a specific substance but those of organic matter in general. In his view, nature comprised only one kind of organic matter—a kind of matter that was common to all organisms, whether animals or plants, and that served not only reproduction but also nutrition and development.[30] That being the case, he described this matter as a "universally prolific substance" or "universal semen."[31] The proof that female animals also possessed this seminal fluid, furnished with the same organic molecules, was crucial to his understanding. By considering Buffon's concept of "universal" organic matter, it becomes possible to grasp the full import of his microscopic studies of the semen. What Buffon observed under the microscope were not living beings, but the first steps in a process of generating life. The initial assemblages of organic molecules that he described in the seminal fluid of both male and female animals illustrated nothing less than the first stage in the reproduction of life—a process that was occurring constantly and everywhere in the whole of the organic world.

The search for a universal principle underlying the generation of life was also pursued by the physician Johann Friedrich Blumenbach in his influential essay *Über den Bildungstrieb und das Zeugungsgeschäfte* (On the formative drive and the business of procreation, 1781).[32] Blumenbach derived his concept of the *Bildungstrieb* (formative drive) from his observations of the regeneration of amputated freshwater polyps and his work on wound healing. Working from these phenomena, he assumed that in "all living creatures, from man to maggot and from cedar to mould," a drive or force was at work that directed not only the process of generation, but also that of reproduction (in the sense of the regeneration of amputated body parts) and nutrition.[33] For Blumenbach, an important feature of the formative drive was its universality: it formed all "organized bodies" to an equal extent, regardless of their length, size and other such physical attributes.[34] In the second edition from 1789, he also emphasized that the formative principle was one of the "forces of life" but that it differed distinctly from all the other forces at work in organized bodies. He regarded the formative drive, in analogy to Newtonian gravity, as being a constantly active power. How the formative drive actually exerted its effect, and the causes (and reasons) of that effect, could not be determined more precisely according to Blumenbach. It was a mysterious force, a *qualitas occulta*.[35]

For Blumenbach generation was one of the three processes subject to this superordinated and generally effective force. Not unlike Buffon, he assumed that the generative matter of both parents, including the "paternal semen," contributed to the generation of a new organism.[36] During fertilization, the "paternal and maternal liquors destined for generation" united and mixed. However, in Blumenbach's view this act and these generative materials were not capable of bringing forth new life on their own. It was only later, when the *Bildungstrieb* took effect, that the hitherto unformed matter began to take form.[37] In other words, Blumenbach's view of generation was characterized by a dichotomy between an immaterial "vital force" and a material devoid of any of the qualities of life. In this respect, Blumenbach's theory of generation differed quite fundamentally from that of Buffon who assumed the existence of vital elements in the material itself. The difference becomes particularly clear in Blumenbach's approach to the spermatic animals. He found equally nonsensical both the idea that spermatic animals were preformed germs of future beings and the attempt to deny their vitality completely: "I cannot conceive how some professed philosophers and natural historians have been led to deny life and voluntary motion of the spermatic animals."[38] The microscopic formations to be found in semen were, in Blumenbach's view, undoubtedly animals, though animals that had no relationship to the actual generative matter. They were located in this physical substance by mere coincidence; he called them "foreign guests of the male semen."[39]

The dualism between an immaterial vital force and unformed material that underlay Blumenbach's theory of generation made it superfluous for him to analyze the organic material more closely, using a microscope to search, as Buffon had, for active material entities that could generate life. In his essay on the formative drive, Blumenbach referred to the parents' "generative substances" or "liquors." For his theory, it was irrelevant to define those substances and the organs that produced them in more detail, since from Blumenbach's perspective, all the formative processes were steered by an immaterial vital force and were not laid down in self-organized organic matter. Yet despite all the differences between Buffon and Blumenbach, their theories share some important common ground. Both conceive of the organic world in its unity and seek to explain the generation of life in ways that can apply across every distinction to all human beings, animals and plants.

ORDERING THE WORLD THROUGH GENDER HIERARCHY: OKEN

About twenty years after the first edition of Blumenbach's theory of the formative drive, Lorenz Oken remarked that the book was to be found "in the hands of every physician and every naturalist."[40] In Oken's view, the extensive and positive reception of Blumenbach's theory was well-deserved. Oken and Blumenbach gave the same reasons for rejecting preexistence theory, especially in its most recent, ovist variant: they both noted that bastards, monsters, and, more generally, resemblances between fathers and their children could not be explained by preexistence theory.[41] Neither Oken nor Blumenbach were in any doubt that "all formation" occurs through epigenesis.[42] Despite this agreement with Blumenbach, Oken's essay *Die Zeugung* (Generation) undertook what he called an "audacious" attempt to write "in a new way" about the origin and generation of life.[43] Oken proposed a theory of generation that separated him not only from Blumenbach but also from Buffon. What was the nature of his innovation?

Oken's criticism was directed at the fact that Blumenbach, while claiming that a confluence of "certain liquors" occurred in generation, failed to address the issue of how to specify this substance and its components more precisely.[44] At the same time, he criticized Buffon's definition of both sexes' generative substance as consisting of fundamentally identical materials. For Oken there was no female semen: "the liquor emitted by the female genitals during intercourse" was, he argued, no generative material and must not be confused with the "real semen." Rather, it was a mucus to lubricate the vagina; "a result of the opening uterine orifice."[45] The woman's contribution to generation was, in Oken's understanding, clearly "the female vesicle." The man, in contrast, contributed a truly generative substance: semen that was uniquely male. Only the male semen contained spermatic animals, which Oken—in contrast to Blumenbach's view—did not consider parasites but rather "essential, indeed the essential element of the entire business of generation."[46] Unlike Buffon, who had fully separated the production of organic molecules from the sexed body, Oken thought that spermatic animals were developed exclusively in the testicles of the male organism.[47] Oken's theory of generation, however, involved a further innovation. As he himself stressed, he contested the views of "most recent physiologists," who considered the female vesicle to be the "central point of epigenesis."[48] According to him, the egg in fact supplied neither "a germ nor elementary organic particles nor anything else material, but merely the form."[49]

Buffon, too, had cast fundamental doubt on the importance of the egg for generation. But whereas Buffon's concern was to formulate a theory based on a symmetrical contribution to generation by both sexes, Oken reintroduced an asymmetry. He again made reference to the Aristotelian dualism between form and matter, but imbued it with a different meaning: instead of contrasting female matter with male immaterial formative power, he postulated the existence of only one material unit—the spermatic animals—that assigned the leading role in generation to the man for two reasons. First, it was the spermatic animals "entering" the "female vesicle" that supplied the raw material for the future embryo. Second, these animals were the driving force of the entire process of generation.[50]

However, in Oken's *Die Zeugung* the significance of the spermatic animals is not exhausted by these two functions. Tellingly, Oken usually referred to the spermatic animals as "infusorians."[51] The term embraced both the microscopic entities observed in the human body and all of the entities that appeared in the course of fermentation, the putrefaction of organic substances, or infusions. Whereas eighteenth-century research into these infusorians attempted to prove the animal nature of the microscopic creatures or to substantiate the hypothesis of spontaneous generation, Oken's objective was a different one.[52] For him, the infusorians were neither animals nor could their emergence be described as the creation of animals that did not exist previously. Instead, he attributed their existence to a process of "coming apart," which referred to the decomposition of a composite organization into its "constituent animals."⁵³ By inverse inference (as he put it), he proposed the hypothesis that all "higher animals" consist of infusorians or "constituent animals." He called them constituent animals or *Urthiere*—"primordial animals"—because they had, like earth, air, and water, come into being at Creation. An additional reason for Oken to consider them primordial animals was that he believed they represented *Urstoffe* or "primordial substances" of the organic "elements in the organic world."⁵⁴ The concept of the infusorians as *Urthiere* was, therefore, connected with the view that they were not merely the fertilizing component of a specifically male generative substance but had a far more comprehensive function: they were, quite simply, the primary units constituting life and generating life.

Although Oken did not use the word *cell* in his 1805 text (at least not with reference to elementary organic forms), historians of biology have often considered his concept of infusorians to anticipate cell theory.⁵⁵ Attempting to identify crucial conceptual developments that led to the emergence of the cell theory during the nineteenth century, François Jacob has highlighted the distinctions between Buffon's organic molecules and Oken's infusorians. Both scholars proceeded from the assumption that living bodies consist of elementary units. But, according to Jacob, whereas Buffon's concept of "organic molecules" was influenced by Newtonian mechanics, the breach with the eighteenth-century's mechanical thinking embodied by Oken was a pivotal move toward the cell theory. At the beginning of the nineteenth century, a living body:

> [C]ould no longer be imagined as a mere association of elements as Maupertuis and Buffon envisaged. Even when Oken again brought up the idea that beings were composed of elements, he did not contemplate autonomous units bracketed together, but units amalgamated in the wholeness of the complete organism. Oken's new idea, from which the cell theory was gradually to emerge, was to consider the bodies of large animals in relation to microscopic beings and to visualize the latter as elements of the former—in short, to imagine the complex living organism as an association of simple living organisms.⁵⁶

Oken's conceptualization of the common basic unit of the organic world in analogy to the smallest living beings was thus, in Jacob's view, a decisive innovation. In contrast to earlier understandings, this unit "could no longer be a simple molecule, an inert element or a portion of matter. It was itself a living body, a complex formation, able to move, feed, and reproduce, a body, in fact, endowed with the principal attributes of life."[57]

I would like to contest Jacob's interpretation by pointing out that even if Buffon did not conceive of his organic molecules as living beings, he certainly did consider them to be mobile, active, and living entities. In Buffon's description of the fundamental unit of the organic world, the term *activity* is of no lesser importance than in Oken. Thus, one of Buffon's conclusions reads: "The life of an animal or vegetable seems to be nothing else than a result of all the particular [activities, all the particular] *lives* (if the expression be admissible) of each of these active particles, whose life is primitive, and perhaps indestructible."[58] A major difference between Buffon and Oken is that Oken reinterpreted this activity, as an exclusively male characteristic. Thus, Oken equates the "active" with the infusorian and the "man," and opposes it to the passive plant or the "woman."[59] The dichotomy of active and passive—in combination with the dichotomy of male and female—however, is not only a key motif in Oken's text, but also appears in the work of other Romantic philosophers of nature, such as Friedrich Joseph Wilhelm Schelling, whose writings had a particularly strong influence on Oken.[60]

This gendered understanding of the "infusoria" is completely left out by Jacob and also by Georges Canguilhem in his essay on the history of the cell theory. Both fail to analyze Oken's book as a whole and focus only on one short quotation from one of the few passages in which gender is not thematized.[61] For Canguilhem, Buffon's and Oken's concepts differ in so far as the latter conceived elementary units of life, which relinquish their individuality to subordinate themselves to a higher unity. Oken's vision of the organism, in Canguilhem's view, results from a rejection of the ideas of the French Revolution and the political philosophy of the Enlightenment: "Oken conceived the organism in the image of society—not society as an association of individuals, as per the political philosophy of the *Aufklärung*, but as the community conceived by Romantic political philosophy."[62] Canguilhem concludes that the history of the cell concept is inextricably entwined with the history of the concept of the individual. In his analysis of vitalism in the Enlightenment era, Peter Hanns Reill likewise locates Oken's theory of generation within the political context of the late eighteenth century. Oken's formative years, Reill reminds us, were dominated by the French Revolution and its consequences. One of Reill's propositions is that in reaction to these political upheavals, which meant chaos and uncertainty for Oken's generation, and especially in opposition to the new ideas of the Enlightenment, Oken and the

Naturphilosophen affirmed a renewal of hierarchies, order, and clarity. However, for Reill, unlike for Canguilhem, Oken's critique of the Enlightenment did not only imply the individual's subordination to the well-being and continuity of society, but also a redefinition of gender relations.[63]

I would like to take up Reill's point and argue that it is perfectly possible to find notions of gender ambiguity and complementarity in the work of Blumenbach and Buffon. For both scholars, however, the ultimate goal was to formulate explanations for the generation of new life that transcended such ambiguities. What characterized Buffon's organic molecules and Blumenbach's formative force was, as I have shown, their status as universal principles: they were equally present and equally effective in all living beings, plants, and animals, regardless of sex and size. In contrast, Oken wanted to validate an order of life that would secure supremacy for one sex—the male. The principle closing Oken's 1805 treatise is "Nullum vivum ex ovo! Omne vivum e vivo." In studies of the history of the cell theory, this principle has been interpreted as a rejection of spontaneous generation and an anticipation of Virchow's famous formula "Omnis cellula a cellula" of 1855.[64] If we read this against the background of Oken's deliberations on gender relations, it becomes clear that, especially with his notion of the *Urthiere*, Oken aimed to disengage the "origin and reproduction of life" completely from the female body.[65]

COMBINING ANTAGONISTIC VISIONS OF THE ORGANIC WORLD

As I have shown in the previous section, Canguilhem and Jacob did not reflect on the central role of dualistic and hierarchical ideas of gender relations in Oken's *Die Zeugung* or on their implications for his notion of organic elementary units. This omission is characteristic for the prevailing view on the history of cell theory. Although numerous studies have discussed the political analogies and metaphors that shaped this theory, it has so far not been analyzed from a gender perspective. In the concluding part of this chapter, I therefore want to pursue my argument by pointing to a few exemplary cases from nineteenth-century sperm and cell research. The first case is Theodor Schwann (1810–1882), who is widely regarded as the "founding father" of cell theory in the late 1830s. Canguilhem and Jacob have related his cell theory to Oken's notion of organic elementary units. Arguing that both Oken and Schwann conceived these elementary units, respectively "cells," in analogy to living beings, Jacob postulates a historical continuity between their concepts.[66] This view disregards an important difference. Whereas Oken understood this unit as a male animal, Schwann explicitly defined the cell in gender-neutral terms as an "individual, an autonomous whole."[67] For Schwann, each cell possessed "an autonomous life" as well as an ability and force to induce organic developmental processes.[68] Whereas for Oken the existence of the organism presupposed the subordination and even the destruction of the individual "infusorians," Schwann regarded the organic whole as the result of the interaction and union of autonomous units. So, if any historical parallel can be drawn, it is between Buffon and Schwann.[69] As I mentioned earlier, Buffon regarded the life of an animal or vegetable as the sum of active particles "whose life is primitive and perhaps indestructible."[70] Both Buffon and Schwann searched for a universal law of organic development, a principle that was not only common for animals and plants, but which could also explain all physiological processes, especially the formation of new organisms. In this context, both proposed a new definition of the egg. Whereas Buffon negated the existence of the egg altogether, Schwann equated the structure and development of the egg to that of other organic tissues.[71] As a result, he did not draw a distinction between the formation of the embryo and other cellular processes in living tissues. Like Buffon, Schwann located the origin of life in a unit that transcended the differences of male and female reproductive substances and bodies. In great similarity to Buffon's theory of reproduction, Schwann's cell theory proposed a vision of the organic world that abstracted from differences between the sexes.

My second example refers to the impact of cell theory for the physiology of reproduction. Historians of biology usually assume that the cell concept was immediately applied to the reproductive process.[72] Yet, the nature and function of the male semen remained a matter of controversy well into the second half of the nineteenth century. An understanding of spermatic animals as parasites largely prevailed until the 1840s,[73] and it was not before the 1850s and 1860s that physiologists began to describe sperms as cells and fertilization as a cellular process.[74] I would like to explain briefly the late arrival of this development by presenting the example of Albert Kölliker (1817–1905). In 1841 Kölliker published a lengthy study on the cellular formation of sperm. Here he compared sperms with other "organic elementary particles," especially blood corpuscles and eggs.[75] At the same time, however, he emphasized the peculiar nature of sperms. Although he did not consider them to be animals, and especially not animals "that come in from outside," he did still see them as having one property reminiscent of their animal nature: their motility. For

Kölliker, this was a feature that fundamentally distinguished sperm from egg. Whereas the "principle of repose" inhered in the egg, the "principle of movement," and therefore a "higher life," was present in the sperm. The union of these two complementary principles was necessary for generation, but as a result of their specific characteristics, Kölliker credited the sperms with the role of initiating the fertilization process.[76] The case of Kölliker represents a hegemonic branch of nineteenth-century physiology of reproduction that was permeated by the gender-dualism of Romantic *Naturphilosophie*. The identification of sperms as a specific form of organic elements that was essentially different from eggs and other cells can be attributed to a dichotomic understanding of gender difference.[77]

Today, sperms and eggs are defined as cells. Yet, fertilization is regarded as a process in which the sperm cell—still referred to by the term *spermatozoon*[78]—"penetrates" the egg. Sperms continue to be described as animals with a "head" and a "tail." Their motility or nonmotility, which—along with their number and shape—is regarded as an indicator of male fertility or infertility and is equated with "vitality" or "death." Buffon's reinterpretation of sperms as initial associations of organic molecules completely overthrew the notion of spermatic animals as living beings. The modern understanding of reproduction seems to owe more to Kölliker's and Oken's views, which highlighted the opposition between the male/active and the female/passive parts.[79] In fact, our perspective on reproduction combines two opposing visions of the organic world. One is based on the principles of universality, individual autonomy, and the functional equality of all organic elements, and one is centered on a hierarchical view of gender difference. In contrast to the former, the latter attributes to the male sex a specific, primary role in the formation of life. The studies and debates discussed in this chapter show how far the thinking about the organic world in the eighteenth and nineteenth centuries was shaped by these two perspectives and the antagonism between them. Moreover, it demonstrates that a gender-based approach allows for a reassessment of well-established narratives in the history of biology.

NOTES

1. See Michel Foucault, *The Order of Things: An Archaeology of the Human Sciences* (New York: Vintage, 1994); François Jacob, *The Logic of Life: A History of Heredity*, trans. Betty E. Spillmann (Princeton: Princeton

University Press, 1993); Wolf Lepenies, *Das Ende der Naturgeschichte: Wandel kultureller Selbstvertsändlichkeiten in den Wissenschaften des 18. und 19. Jahrhunderts* (München: Hanser, 1976).
2. This discovery is well known and has often been discussed, see Jean Rostand, *La formation de l'être: Histoire et idées sur la génération* (Paris: Hachette, 1930), 79–88; Jacques Roger, *Les sciences de la vie dans la pensée française du XVIIIe siècle* (Paris: Albin Michel, 1993), 293–322; Jörg Jantzen, "Theorien der Reproduktion und Regeneration," in Friedrich Wilhelm Joseph Schelling, *Ergänzungsband zu Werke Band 5 bis 9: Wissenschaftshistorischer Bericht zu Schellings naturphilosophischen Schriften 1797–1800* (Stuttgart: Frommann und Holzboog, 1994), 588–595; Carlo Castellani, "Spermatozoa Biology from Leeuwenhoek to Spallanzani," *Journal of the History of Biology* 6 (1973):37–68; John Farley, *Gametes & Spores: Ideas about Sexual Reproduction* (Baltimore: Johns Hopkins University Press, 1982), 17–23; *Antoni van Leeuwenhoek 1632–172: Studies On the Life and Work of the Delft Scientist Commemorating the 350th Anniversary of His Birthday*, ed. L. C. Palm and H. A. M. Snelders (Amsterdam: Rodopi, 1982); Edward Ruestow, "Images and Ideas: Leeuwenhoek's Perception of the Spermatozoa," *Journal of the History of Biology* 16 (1983):185–224; Carla Pinto-Correia, *The Ovary of Eve: Egg and Sperm and Preformation* (Chicago: University of Chicago Press, 1997).
3. On the formation and diffusion of the preformation theory in the seventeenth and eighteenth century, see Roger, *Les Sciences*, 324–439; Shirley A. Roe, *Matter, Life, and Generation. Eighteenth-Century Embryology and the Haller-Wolff Debate* (Cambridge: Cambridge University Press, 1981); Pinto-Correia, *The Ovary of Eve*.
4. See Georges-Louis Leclerc, Comte de Buffon, *Histoire naturelle, générale et particulière* (Paris: Imprimerie Royale, 1749–1788), vol. 2:168. The translation of the following quotations by Buffon partly follows the English edition: Georges-Louis Leclerc de Buffon, *Natural History, General and Particular*, trans. William Smellie, vol. 2 (London: Strahan and Cardell, 1781).
5. Buffon, *Histoire naturelle*, vol. 2:54.
6. Ibid., 168. On Buffon's concept of organic molecules, see Roger, *Les Sciences*, 542–558; Peter Hanns Reill, *Vitalizing Nature in the Enlightenment* (Berkeley: University of California Press, 2005), 33–70.
7. Buffon, *Histoire Naturelle*, vol. 2:172.
8. Ibid., 73–174.
9. Ibid., 178–185 and 241–254.

10. Ibid., 186.
11. Ibid., 189 and 189–201.
12. Ibid., 169.
13. On Harvey and de Graaf, see Jantzen, "Theorien der Reproduktion," 566–573; Pinto-Correia, *The Ovary of Eve*, 42–45; Thomas Laqueur, *Making Sex: Body and Gender from the Greeks to Freud* (Cambridge, MA: Harvard University Press, 1990), 142–148, 182.
14. See Buffon, *Histoire Naturelle*, vol. 2:99 and 288–292.
15. Ibid., 130, 133, 288, and 292–297.
16. Ibid., 86–90. On Aristoteles' notion of semen and generation, see Gianna Pomata, "Vollkommen oder verdorben? Der männliche Samen im frühzeitlichen Europa," *L'Homme* 6 (1995):59–85.
17. Buffon, *Histoire Naturelle*, vol. 2:77–68 and 158.
18. Ibid., 92–97. On Galen's two-seed theory, see Pomata, "Vollkommen," 166–170.
19. On Maupertuis and his influence on Buffon, see Roselyne Rey, "Génération et Hérédité au 18e siècle," in *L'Ordre des Caractères: Aspects de L'hérédité dans L'histoire des Sciences de L'homme*, ed. Jean-Louis Fischer and Claude Bénichou (Paris: Sciences en Situation, 1989), 7–41; Rostand, *La Formation*, 103–107.
20. Buffon, *Histoire Naturelle*, vol. 2:202–203.
21. Ibid., 208.
22. Ibid., 210–221.
23. Ibid., 169–170.
24. Ibid., 329.
25. Ibid., 330–331. He assumed that the "assemblage" of an embryo from organic molecules occurred according to a formative principle called "internal mould": "c'est de la réunion de ces parties organiques, renvoyées de toutes les parties du corps de l'animal ou du végétal, que se fait la reproduction, toujours semblable à l'animal ou au végétal dans lequel elle s'opère, parce que la réunion de ces parties organiques ne peut se faire qu'au moyen du moule intérieur, c'est-à-dire, dans l'ordre que produit la forme du corps de l'animal ou du végétal, & c'est en quoi consiste l'essence de l'unité & la continuité des espèces." Buffon, *Histoire Naturelle*, vol. 2:258. On Buffon's generation theory, see Roger, *Les Sciences*, 542–558; Jantzen, "Theorien der Reproduktion," 606–609; Rostand, *La Formation*, 112–117 and Susanne Lettow in this volume.
26. See Buffon, *Histoire Naturelle*, vol. 2:336–337.
27. Ibid., 72.

28. Ibid., 301.
29. Ibid., 422 and 280.
30. Ibid., 420.
31. Ibid., 306, 304, and 425.
32. On Blumenbach's theory of the *Bildungstrieb* and its influence on Kant and the naturalists and physiologists of the late eighteenth and early nineteenth century, see Robert J. Richards, *The Romantic Conception of Life: Science and Philosophy in the Age of Goethe* (Chicago: University of Chicago Press, 2002), 207–237; Reill, *Vitalizing Nature*, 166–171; Jantzen, "Theorien der Reproduktion," 636–668; Timothy Lenoir, "Kant, Blumenbach, and Vital Materialism in German Biology," *Isis* 71 (1980):77–108; Timothy Lenoir, *The Strategy of Life: Teleology and Mechanics in Nineteenth-Century German Biology* (Chicago: University of Chicago Press, 1982).
33. Johann Friedrich Blumenbach, *Über den Bildungstrieb und das Zeugungsgeschäfte* (Stuttgart: Fischer, 1971), 9–13. The following quotations by Blumenbach partly follow the translation by Alexander Crichton, J. F. Blumenbach, *An Essay on Generation* (London: Cadell, 1792).
34. Blumenbach, *Über den Bildungstrieb und das Zeugungsgeschäfte*, 55 and 85.
35. Johann Friedrich Blumenbach, *Über den Bildungstrieb* (Göttingen: Dietrich 1789), 25–26.
36. Blumenbach, *Über den Bildungstrieb und das Zeugungsgeschäfte*, 20. Blumenbach had already pointed to the role of semen in his *Handbuch der Naturgeschichte* (1779); see Lenoir, "Kant, Blumenbach, and Vital Materialism," 82; Richards, "Kant and Blumenbach," 17. In this context he referred to the plant hybridization experiments of Joseph Gottlieb Koelreuter (1733–1806), see Staffan Müller-Wille, Vitezslav Orel, "From Linnean Species to Mendelian Factors: Elements of Hybridism, 1751–1870," *Annals of Science* 64 (2007):171–215, 182–191.
37. Blumenbach, *Über den Bildungstrieb und das Zeugungsgeschäfte*, 42 and 46.
38. Ibid., 32.
39. Ibid.
40. Lorenz Oken, *Die Zeugung* (Würzburg: Goebhardt, 1805), 102.
41. Ibid., 37–57.
42. Ibid., 107
43. Ibid., iii.
44. Ibid., 108.

45. Ibid., 98.
46. Ibid., 101 and 102.
47. Ibid., 61.
48. Ibid., 101.
49. Ibid., 103.
50. Ibid.
51. The term "infusorians" was introduced by Martin Frobenius Ledermüller in 1763. See *Geschichte der Biologie: Theorien, Methoden, Institutionen, Kurzbiographien*, ed. Ilse Jahn (Jena: Fischer, 1998), 267. It referred to all microscopic animals that were up to than subsumed under the Latin notion "animalculae." The study of infusorians started in the late seventeenth century; see Marc J. Ratcliff, *The Quest for the Invisible: Microscopy in the Enlightenment* (Farnham: Ashgate, 2009), 177–215.
52. On the history of spontaneous generation, see John Farley, *The Spontaneous Generation Controversy from Descartes to Oparin* (Baltimore: Johns Hopkins University Press, 1977).
53. Oken, *Die Zeugung*, 21.
54. Ibid., 22.
55. See Georges Canguilhem, *Knowledge of Life*, trans. Paola Marrati and Todd Meyers (Fordham University Press 2008), 39–40; Jacob, *The Logic of Life*, 114–116; William Colemann *Biology in the Nineteenth Century: Problems of Form, Function, and Transformation* (New York: Wiley, 1971), 25–26; Henry Harris, *The Cells of the Body: A History of Somatic Cell Genetics* (Cold Spring Harbor: Laboratory Press, 1995), 1–3; Jahn, *Geschichte der Biologie*, 290–292. Ilse Jahn points out that Oken used the term *cell* as early as 1809, see p. 292.
56. François Jacob, *The Logic of Life*, 114–115.
57. Ibid., 116.
58. Buffon, *Histoire Naturelle*, vol. 2:340.
59. Oken, *Die Zeugung*, 116 and 150.
60. On gender dichotomies in Schelling's philosophy see Lettow and Alison Stone in this volume, and Claudia Honegger, *Die Ordnung der Geschlechter: Die Wissenschaften vom Menschen und das Weib* (Frankfurt am Main: Campus, 1991), 182–190. Oken himself referred to Schelling's influence on his thinking, see Oken, *Die Zeugung*, iv. On the general importance of gender differences in the late eighteenth century, see Ute Frevert, *Mann und Weib, und Weib und Mann: Geschlechter-Differenzen in der Moderne* (München: Beck, 1995); Honegger, *Die Ordnung*; Laqueur, *Making Sex*.
61. "Die Verbindung der Urthiere im Fleische ist nicht zu denken, als etwa

eine mechanische Aneinanderklebung eines Thierchens an das andere, wie ein Haufen Sand, in dem keine andere Vereinigung stattfindet, als des Beieinanderliegens mehrerer Körnchen—nein! Ähnlich dem Verschwinden des Wasserstoffes und Sauerstoffs im Wasser, des Quecksilbers und Schwefels im Ainober, ist es eine wahre Durchdringung, Verwachsung, ein Einswerden all dieser Thierchen, die von nun an kein eigenes Leben führen, sondern alle, im Dienste des höheren Organismus befangen, zu einer und derselben Funktion hinarbeiten, oder diese Funktion durch ihr Indentischwerden selbst sind. Hier wird keines Individualität geschont, dies geht für sich schlechthin zu Grunde, und aber nur uneigentlich gesprochen, die Individualität aller bilden nur Eine Individualität—jene werden vernichtet, und diese tritt erst aus jener Vernichtung hervor." Oken, *Die Zeugung*, 23.
62. See Canguilhem, *Knowledge of Life*, 39–42.
63. See Reill in this volume and Reill, *Vitalizing Nature*, 220–236. On the issue of gender differences in Oken's theory of generation, see also Dietmar Schmidt, "Klimazonen des Geschlechts: Zeugung um 1800," *Metis* 17 (2000):5–29.
64. See Harris, *The Cells of the Body*, 3.
65. Oken, *Die Zeugung*, iii.
66. See in particular Renato Mazzolini, *Politisch-biologische Analogien im Frühwerk Rudolf Virchows* (Marburg: Basiliken-Presse, 1988); Laura Otis, *Membranes: Metaphors of Invasion in Nineteenth-Century Literature, Science and Politics* (Baltimore: Johns Hopkins University Press, 1999); Eva Johach, *Krebszellen und Zellenstaat: Zur medizinischen und politischen Metaphorik in Rudolf Virchows Zellularpathologie* (Freiburg i. Br. u.a.: Rombach, 2008).
67. Theodor Schwann, *Mikroskopische Untersuchungen über die Übereinstimmung in der Struktur und dem Wachstume der Tiere und Pflanzen* (Leipzig: Engelmann, [1839] 1910), 4.
68. Schwann, *Mikroskopische Untersuchungen*, 189.
69. Buffon's place in the history of the cell theory has often been discussed, but not from a gender perspective, see Canguilhem, *Knowledge of Life*, 33–37; Staffan Müller-Wille, "Cell theory, specificity, and reproduction, 1837–1870," *Studies in History and Philosophy of Biological and Biomedical Sciences* 41 (2010):225–231; Daniel J. Nicholson, "Biological atomism and cell theory," *Studies in History and Philosophy of Biological and Biomedical Sciences* 41 (2010):202–211.
70. See Buffon, *Histoire naturelle*, vol. 2:340.

71. See Schwann, *Mikroskopische Untersuchungen*, 40–60.
72. See, for instance Jacob, *The Logic of Life*, 125.
73. See, for instance, the entry "Vom Samen" in Johannes Müller, *Handbuch der Physiologie des Menschen* (Coblenz: Hölscher, 1840), vol. 2:633–651.
74. See Ferdinand Keber, *Ueber den Eintritt der Samenzellen in das Ei: Ein Beitrag zur Physiologie der Zeugung* (Königsberg: Bornträger 1853); Schweigger-Seidel, "Über die Samenkörperchen und ihre Entwicklung," *Archiv für mikroskopische Anatomie* 1 (1865):309–335; Adolph Lavalette St. George, "Über die Genese der Samenkörper," *Archiv für mikroskopische Anatomie* 1 (1865):403–414.
75. Albert Kölliker, *Beiträge zur Kenntniss der Geschlechterverhältnisse und der Samenflüssikeit wirbelloser Thiere nebst einem Versuch über das Wesen und die Bedeutung der sogenannten Samenthiere* (Berlin: Logier, 1841), 75.
76. Kölliker, *Beiträge zur Kenntniss der Geschlechterverhältnisse*, 74 and 82, see also 83–84.
77. For a longer development of this argument, see Florence Vienne, "Vom Samentier zur Samenzelle: Die Neudeutung der Zeugung im 19. Jahrhundert," *Berichte zur Wissenschaftsgeschichte* 32 (2009):221–229.
78. It was Karl Ernst von Baer (1792–1876), one of the pioneers of modern embryology, who introduced the modern name "spermatozoa" in 1827, yet even he considered these to be parasites and denied them any role in generation. See Karl Ernst von Baer, "Beiträge zur Kenntnis der niederen Tiere," *Verhandlungen der kaiserlichen Leopoldinisch Carolinischen Akademie der Naturforscher* 13 (1827):558–659, p. 640; and the *Die Physiologie als Erfahrungswissenschaft*, ed. Friedrich Burdach (Leipzig: Voss, 1826), vol. 1:90–95.
79. Helga Satzinger shows that this opposition still informed Theodor Boveri's theory of fertilization, see Helga Satzinger, *Differenz und Vererbung: Geschlechterordnungen in der Genetik und Hormonforschung 1890–1950* (Köln: Böhlau, 2009), 83–111.

3

THE SCIENTIFIC CONSTRUCTION OF GENDER AND GENERATION IN THE GERMAN LATE ENLIGHTENMENT AND IN GERMAN ROMANTIC *NATURPHILOSOPHIE*

PETER HANNS REILL

In many ways, traditional accounts of science have cast it as the triumph of impartial, objective understanding over pure subjectivity. However, with the rise of postmodernism, this view has been severely attacked, leading many postmodernists to see all so-called objective statements as inherently subjective—part of the power/knowledge dyad Foucault so powerfully analyzed. In the debates that have arisen in the past forty years, one topic, in which the tensions between so-called objective knowledge and subjective interest-driven concerns has played a crucial role, has been the analysis of modern gender construction, often focusing on the period from the late Enlightenment to early romanticism. In such diverse works as Thomas Laqueur's *Making Sex* and Claudia Honegger's *Die Ordnung der Geschlechter*,[1] two important moments are traced that supposedly took place in the late Enlightenment and early romanticism that reified prejudicial stereotypes of male and female qualities into normal scientific discourse. The first was what Laqueur called the creation of the two-sex model, which assumed an essential biological difference between males and females; the second was what Honegger describes as "biologizing" this model, giving these gender distinctions the stamp of scientific objectivity. For both authors, these major steps in objectifying gender inequality took place in the late Enlightenment. In what follows, I question this assumption and seek to offer a more complex interpretation of gender construction in this period by differentiating between Enlightenment and romantic gender constructions.

In the eighteenth-century, the "scientific" discussion of gender was part of the larger discussion of generation, which occupied an important place in the intellectual economy of late Enlightenment thinkers.[2] It mobilized a host of cultural, social, and political symbols and metaphors that overdetermined its importance for late Enlightenment culture. When the issue first came to a head in the mid-century, the central disputes focused on how creation took place, what defined a species, how one could represent change in and through time, how one defines a human within the empirical world the human inhabits, to what extent hierarchies were natural and unchanging, and what the connection was between traditional Christian concepts and natural philosophy. In this phase of the discussion, generation and gender were only vaguely linked, and when done, gender played not a big part in the drama that captured the imagination of educated Europeans.

Over the course of the last half of the century, however, gender—often interpreted in terms of sexual difference—increasingly occupied a more important place within the parameters of the competing discourses of generation. Rather than focus solely on the process of either preformation or epigenesis, one increasingly began to investigate the actors in this play—namely, the male and the female, and asked what separated and linked them. As the question was posed, the answers tended to lead to a differentiation between both sexes.

In this sense, Thomas Laqueur's assertion that the eighteenth century invented the modern idea of sex, which served "as a new foundation for gender" is correct.[3] But though gender differentiations increasingly were expressed in the language of sexuality, they were hardly ever associated solely with sexuality or with specific sexual organs as defining instruments for the sex in question. Though sexuality increasingly became the object of scientific inquiry, the differentiation between the two sexes according to physical difference still was not seen as the primary determinant in gender relations. The biologizing of sex, the establishment of incommensurable differences between male and female, and the distinction between "the natural and the social"[4] in the sexual sphere could only take place when nature was "de-moralized," separated from morality and juxtaposed to it. In most parts of Europe and the Americas, this did not occur until the mid-nineteenth century and was not a necessary consequence of what took place in the Enlightenment. Certainly, most late

Enlightenment thinkers did not draw a radical separation between the sexes, and if they did, it usually was not based on biology.

This was especially true for the vast majority of Enlightenment vitalists, such as Georges-Louis Leclerc, Comte de Buffon, and Johann Friedrich Blumenbach, one of the leading German vitalist life scientists of the period. As Florence Vienne makes clear in her chapter in this volume, vitalists such as Buffon and Blumenbach adopted a Hippocratic explanation for generation that located the causes of sexual differentiation in the relative powers of the mixing of "male and female fluids," or powers, for determining the sex of the new individual and in shaping its particular features. Enlightenment vitalists assumed that the appearance of sexual differentiation also signaled a "moral" differentiation between the genders. But most vitalists modified such distinctions by subsuming sexual and gender difference within the unity of the human species. This imperative, along with the assumption that all humans were mixtures of masculine and feminine qualities, led to constructions of gender that tended more toward idealizations of androgyny than to incommensurable differentiation.

If one looks at the gendered world inhabited by many educated men and women during the late eighteenth century, this tendency toward androgynous attitudes becomes evident. Let me cite two examples. The first was the widespread shift in forms of address, expressions of emotion, and polite conventions between people of both sexes. The second, partly arising as a consequence of the first, was a definite loosening of the taboos concerning male and female social and sexual conventions, which led many to question the traditionally accepted models of gender that had been forged during the early modern period and to propose new alternatives to them.

With respect to the first point, there is no doubt that the last third of the century witnessed an intensification of emotion between individuals expressed in terms of intimacy with very strong erotic overtones, suffused with ideas of love. This is evident if one looks at the correspondence between friends during this period. The hetero- and homoerotic flavor of these letters is unmistakable, though their actual meaning is too often taken today at face value, that is, as signaling more than they were. Expressions of love and affection were used as frequently between members of the same sex as between those of the opposite sex and often by the same writers employing the same terms for either sex. Males addressed males as they addressed females in a language that, for earlier generations, would have seemed highly inappropriate and for later generations too intimate. The letters suggest a breakdown of normal gender relations, showing a narrowing of the distance between the genders, expressed in an eroticized language.

This highly charged erotic-emotional universe also witnessed the emergence of a group of impressive women whose personality and learning placed them on an equal footing with the men with whom they met, socialized, corresponded, married, or had liaisons. These women provided physical proof for the general questioning of traditional gender relations. From Madame de Staël and Mary Wollstonecraft through the Berlin *salonnières*, such as Henriette Herz and Rahel Varnhagen, to the daughters of famous philosophers and scholars, such as Brendel Veit, Therese Heyne, Caroline Michaelis, and Dorothea Schlözer, women such as these played an important role in shaping the cultural and emotional life of those growing up in the late Enlightenment. Taken together, the interactions between and among the sexes and the highly charged emotional environment in which they were played out appeared to seriously threaten the traditional rules governing gender relations.

One of the most powerful ideas that evolved within this shifting world where ambiguity appeared to reign, was that the new person, whether female or male, should incorporate, as best as possible, the positive aspects of the other gender. Thus, categories of male and female, though increasingly constructed according to certain contrasting typologies, were seen as both the limits within which life was lived and also as one-sided exaggerations, which should be modified through interaction. The Enlightenment vitalists proposed a reciprocity and mutual interchange between masculine and feminine qualities—the first representing activity, the other receptivity: an ideal recently characterized by Anne-Charlott Trepp as "tender masculinity and self-reliant femininity" (*sanfte Männlichkeit und selbstständige Weiblichkeit*).[5]

AMBIGUITY AND INTERACTION: WILHELM VON HUMBOLDT

Wilhelm von Humboldt, a major figure of the late Enlightenment, student of Blumenbach, friend of Georg Forster, Goethe, Schiller and closely involved with the intellectual elites of Germany, France, England, and Italy, presents us with a prime example of how these ideas were played out in his essay *Ueber den Geschlechtsunterschied und dessen Einfluss auf die organische Natur* (On gender difference and its influence on organic nature). It appeared in Schiller's enormously important periodical *Die Horen* in 1795. This essay serves as a clear expression of late Enlightenment thinkers' desire to differentiate yet merge gender categories and, more importantly, of the distance separating their vision of gender from that of later romantic *Naturphilosophie*.

Humboldt's essay was an ambitious attempt to interpret the process of artistic and poetic creation using categories drawn from generation in the natural realm. His desire was to go well beyond the "limited sphere," in which the aesthetic discussion had been held and "to transfer it to an immeasurable field."[6] He undertook this project because he believed "physical nature and moral nature formed a great unity and the appearances in both accord to the same laws."[7] Humboldt took his model of natural generation directly from Blumenbach and the Enlightenment vitalists. Life was the instantaneous product of the interaction between living forces. It began with a living point of matter, and through the action of the *Bildungstrieb* was formed, maintained, and compensated for when damaged. Thus life was based on two necessary components—matter and formative force acting on and through it. The act of generation resulted from the interaction between masculine and feminine forces, both of them active, yet both working in different directions. Male forces were spontaneous; female ones receptive. Nature, forever alive, progressed through the continual interaction of these two forces, which insured the continuation of the species, led to its progress, and constituted nature's unity.[8]

What applied to generation in the physical world was, Humboldt asserted, also true for the spiritual. All creation had to be understood as the result of the active interchange between energetic masculine and feminine forces. As in the physical world, each initially worked in one direction. The masculine force was more prone to action; the feminine, more to reaction. Masculine forces demonstrated more spontaneity: they tended to dissect, to destroy, to isolate. They were more likely to be expressions of power. In the intellectual sphere, masculine forces were analytic, led by cold reason. Feminine forces were receptive: they tended to assimilate, to collect, to join. They were more likely to be expressions of profusion (*Fülle*). In the intellectual sphere they were synthetic, guided by warm fantasy. Thus Humboldt constructed a series of dyads that expressed the differences between male and female forces. The most prominent were: active/receptive, reason/imagination, cold/warm, hard/soft, and analytic/synthetic. These were supplemented by associating each power with types of knowledge. The male principle was more predominant in the search for evident truth; the female in the search for beauty. In the realm of knowledge creation, philosophy represented the masculine drives, while history typifies the feminine.

These characterizations already seem to announce the gender stereotypes that later were associated with the biologizing of sexuality with their direct linkage of physical attributes to psychological categories. And Humboldt is sometimes interpreted as one who contributed to forging this new vision of gender founded on sexuality. Yet a careful reading of the work, when located within the parameters of Enlightenment vitalism, demonstrates, I believe, the opposite.

One of the most important differences between Humboldt's vision and later gender constructions was that Humboldt was very wary of equating gender qualities with the physical. Only in a few places did he make direct reference to the physical differences between men and women. At times he said women more closely embodied the idea of beauty and in one section he remarked: "In the observation of nature, even a quick glance will see in the masculine sex more an expression of power than in the feminine," and in the feminine, "more an expression of fullness." But he also modified this comparison by saying such differences were at best relative, not absolute.[9] But, he even went further than relativizing the connections between the physical and the moral. He argued that neither one of these distinctions could ever exist as an isolated thing in and for itself. Every active force was receptive, every receptive force active, for "every pure separation contradicts the analogy of the laws of nature."[10]

This observation underscores a second important element of Humboldt's analysis. His distinction between masculine and feminine was not based on the dyad of active versus passive—which became the central category for later formulations—but rather on a differentiation founded on two equally active forces, working, as he said, in different directions. The difference between them did not rest in ability or degree of activity. "In humans," he argued,

> [S]pontaneity and receptivity reciprocally correspond to one another. The spontaneous spirit is also the most sensitive; the heart that is most receptive for every impression returns each with the most active energy. Thus, only the different direction distinguishes masculine from feminine power. The first, because of its spontaneity, begins with action, but then because of its receptivity absorbs the counteraction. The second operates in the opposite direction. It absorbs the thrust and returns it with spontaneity.[11]

In this construction, strong male forces required strong female forces for generation to be successful. Humboldt's theoretical assumptions were confirmed by his personal affinities. He was attracted to strong females, whether they were intellectually powerful women, such as Henriette Herz, Caroline Dacheröden (his wife), Therese Heyne, and Madame de Staël, or physically strong females, such as the ferry woman who awakened in him powerful sadomasochistic sexual fantasies. "Between Duisburg and Crefeld one crosses the Rhine with a ferry. On the ferry, amongst the workers was a young woman, outwardly ugly, but strong, masculine, industrious. It is inconceivable how alluring such a sight of active physical power in women—especially from the lower orders—is for me."[12] Strong, energetic females and female forces were central to nature's plan.

Humboldt's distinction between masculine and feminine forces was, therefore, not designed as a hierarchy in which the male took prime place. Each force incorporated an element necessary to the drama of life and was of equal worth. This was made clear in an important footnote Humboldt added in the essay when talking about creation and genius. There were works of genius, he argued, in which masculine reason took precedence over female fantasy and works in which female fantasy and fullness were predominant. To illustrate this point, he presented in the footnote four sets of pairs, in which the first of each was "female," the second "male." They were: "*Homer* and *Virgil, Ariosto* and *Dante, Thompson* and *Young, Plato* and *Aristotle*."[13] Given Humboldt's tastes, this list did not demonstrate a preference for one or the other categories. He preferred, for example, Homer to Virgil and Aristotle to Plato. This lack of preference was also illustrated by the way Humboldt once characterized himself in a fragment of a short autobiographical sketch composed in 1816. The description places Humboldt very much in the female camp, though later self-descriptions did the opposite. In the fragment, he talked about himself being "pregnant with ideas" that were never brought to fruition. He declared his intellectual goal to be "to apprehend the world in its individuality and its totality," and complained of his inability to strictly separate "the individual from the general." It was the "fullness" (*Fülle*) of ideas that attracted him.[14] The contradictory self-evaluations Humboldt produced over his lifetime testify to the ambiguous blending of gender categories characteristic of Humboldt and many of his contemporaries such as Goethe, Christian Gottfried Körner, Friedrich Gentz—a friend and later political opponent—and his intellectual adversary, Friedrich Schlegel.

In the essay's footnote, comparing the gendered products of genius, Humboldt added another example—one that points to the central message of the essay. The two major characteristics Humboldt cited as separating the basic pairings were: warm, sensuous fantasy for the female group, and cold, hard reason for the males. There was, however, a third possibility, one that mediated between the hardness of reason and the fullness of imagination. Humboldt's idea here bears a close structural resemblance to Schiller's concept—elaborated in his *Über die Ästhetische Erziehung des Menschen* (On the aesthetic education of man, 1794)—of the mediation between the *Stofftrieb* and the *Formtrieb* effected by the *Spieltrieb*, but infused with a gendered dimension missing in Schiller's work. As an example, Humboldt cited Sophocles, who stood midway between Aeschylus and Euripides. The image of androgynous mediation between the two powers runs throughout the whole essay. Not only were the two energies to be maximized individually, it was necessary to have them interact and merge in a higher union. "For only by combining the characteristics of the two sexes can perfection (*das Vollendete*) be generated."[15] Failure to join them could lead to destructive anarchy on one side, or to stifling stasis on the other. If one-sidedness was to be avoided, then the male must bind his natural spontaneity to a strong law; the female, who represents the feeling of law, must animate herself from within through spontaneity. In short, Humboldt used the metaphors of merging, marriage, and mediation to describe how a higher harmony between individual male and female forces could be achieved.

Humboldt's construction of gender not only pertained to individuals. It was embedded within a larger historical and theoretical framework, which proclaimed the vitalist belief in purposive change effected through free activity. It was supplemented by ideas drawn from Schiller and Kant about the tasks of reason and the vocation of the human race. But Humboldt added his own, highly original spin on these ideas by focusing on gender. Humboldt believed that only through the intensification and combination of male and female principles could the human attain the "ideal which reason has prescribed." The reconstruction of gender then was the necessary condition for a leap in human consciousness and its progressive development. Humboldt expressed this goal utilizing an eroticized language. "This elevated vocation can only be established when the effects resulting from [the two forces operating freely] are entwined around each other; the inclination, which leads one to longingly approach the other, is *love*."[16]

PURE MASCULINITY, SEPARATION, AND POLARIZATION: LORENZ OKEN

If one turns directly from Humboldt's essay on sexual differentiation to *Die Zeugung*, written by Lorenz Oken, a leading German *Naturphilosoph* in 1805, it might appear as if they had been produced in two radically different eras.

Yet only ten years separated their publication. They offer striking proof of the idea that no age can be compartmentalized under simple rubrics. Though Humboldt was only twelve years older than Oken, their formative years were very different. Oken, as a young man, was thrust into the world of the Napoleonic wars, the defeat of Prussia, and the failure of the French Revolution to achieve the ideals to which most Germans had at first been attracted. The ambiguities of life and nature, which men and women of the late Enlightenment accepted as the announcement of possibilities for the construction of new social relations, appeared to many of Oken's generation as dangerous, leading to uncertainty and chaos.

The *Naturphilosophen*'s desire to attain certainty clearly spilled over into the world of gender relations. In fact they may have become concentrated there, revealing in a purer light the yearning for order, clarity, and hierarchy that, I believe, directed the *naturphilosophic* project.[17] The ambiguous and highly charged sexual atmosphere of the late eighteenth century, in which traditional boundaries were breached, became too difficult for many to navigate or to comprehend. Even the first generation of *Naturphilosophen*, people such as the Schlegel brothers and Schelling, who were active participants in this general scrambling for gender norms during the early stages of the French Revolution, pulled back once successfully married and established. Ambiguity, the blurring of boundaries and the transmutation of gender categories became associated with the negative effects of what was defined as the worldview that supposedly had spawned the excesses of the Revolution. The imperative for many was to return to older gender norms but to do so using the instruments of the latest natural philosophical reasoning—to establish new criteria on which the traditional norms could be revivified, supported by "scientific" insight.

Oken's text indicates the manner in which this reformulation of gender was carried out. He considered it his task to work out general ideas proposed by Schelling. Schelling had argued that sexual differentiation was a universal principle in nature, found everywhere. "Absolute sexlessness is nowhere to be demonstrated in nature, and a regulative principle demands, *a priori*, to assume sexual differentiation everywhere in organic nature."[18] Oken wrote *Die Zeugung* "Generation" to demonstrate the validity of the general principles of Schelling's *Naturphilosophie* when applied to gender and generation, for Schelling, as Alison Stone argues, "had not elaborated on the nature of the two opposed sexes," but only tacitly pointed to their separation.[19]

Oken, following up on Schelling's concept of radical sexual differentiation, also sought to resolve all of the controversies that had arisen between preformationists, epigenesists, and supporters of spontaneous generation (*generatio aequivoca*) as Vienne makes clear in her chapter in this volume. Oken's argument was based on the familiar *naturphilosophic* principles that creation was spontaneous, universal, and made manifest in the appearance of a polar duality within a *trias*. Using the language of progression (though denying its reality), Oken argued that the first appearance of living matter was in the form of a primal-matter (*Urtstoff*), represented by what he called an infusions-animalcule (*Infusionsthier*) or a polyp. This primal-matter represented the positive masculine principle and therefore the radial line that extended to infinity.[20] It constituted the essence of all creation, found everywhere and always in the same quantity, though in different forms.

All creation consisted in combinations of primal-material and their disintegrations. Nothing new or different could be created after the original appearance of reality. Thus all forms of matter were various expressions of primal-matter, either joined or reduced to its more basic essential appearance. But the joining of primal-matter, or the infusions-animalcules, in a more complex unity could not be explained mechanically, a position central to Schelling's philosophy. Further, Oken's explanation reveals the hierarchical nature of his thought, whereby individuals must subordinate themselves to the whole. The combination of essential matter into different forms was,

> [S]imilar to the disappearance of hydrogen and oxygen in water. . . . It is a true permeation, a coalescing, a becoming one [*einswerden*] of all of these animalcules, which from then on lead no independent life, instead all are subordinated in the service of the higher organism.[21]

For this reason, all creation happened at once: "with the act of impregnation . . . the total embryo arises at once with one blow . . . ; it is not formed over time, which also means that one is not allowed to speak of a formation before impregnation."[22] In this formulation, Oken denied both preformation and epigenesis and opted for a modified interpretation of spontaneous generation in which primal-matter went through instantaneous transformations or polarizations. Since primal-matter constituted all material forms and was universally distributed, it was, by definition, sexless. It could not be otherwise, for if it had sex, all products of creation would have to be males. Pure masculinity was not tainted by the appearance of sex, though all living forms arose through the male principle. What applied to all of nature, applied equally to humanity: "The total human race is nothing but a continually propagating male."²³

But, propagation in animals could not occur without the appearance of sex. Oken argued, following Schelling, that sex arose as the necessary negation of the original sex-free male affirmation. It contained an inner contradiction because every negation included the affirmation, which it negated. The negative form was a hybrid (*Zwitter*)—the essential definition of universal sexuality. If the affirmation was male, then the negation was female. It was embodied in the plant, the absolute opposite of the infusions-animalcule or polyp.

> The character of the plant is thus an inner duality. [. . .] This dissatisfaction of the plant with itself, this striving to seek a complement outside of itself on which it could hang, reveals itself as an internal duplicity, not in the numerical sense, but as a double sidedness, as polarity. This duality is sex—the essential character of the plant is sex. It is thoroughly sex in the sense that it is a plant.²⁴

The plant negation of the polyp affirmation drew the radial line back into itself, forming a circle. Thus the essence of femininity was embodied in the circle, the pustule (*Bläschen*), or, in animals, the womb where spermatic material was collected and polarized.

The synthesis of polyp and plant was the animal—the polyp serving as the active principle and the plant as the passive. Combined, they united line and circle to form an ellipse. "The animal is the highest union of the polyp and the plant, the line and the circle; the fusion of the two in one produces the ellipse. . . . The binding idea of animality is tied to this idea, not the circle, not the line, but both in this idea, the most beautiful in form and harmony."²⁵ Oken's choice of the ellipse as the synthetic form uniting circle and line again reveals the abstract, timeless nature of *naturphilosophic* thought; the ellipse is static and one-dimensional, a sharp contrast to Goethe's attempt to visually represent the interaction of line and circle over time as the spiral.

This uniting of polyp and plant in the animal was not a merging or exchange between the two principles as Humboldt and the Enlightenment vitalists would have described it. According to Oken, such an intermingling was impossible. "Nothing, whether it be magnitude or matter can change over to its absolute opposite."²⁶ The separation between male and female was radical, with only one proviso. Though separated, the female as negation was dependent on the male and, in fact, was defined in and through the male.

> The ruling character of the animal in animality is separation of the sexes in two individuals. The masculine sex is independent from the female, because it is an infusions-animalcule. It is the first organism in nature. The female sex is dependent upon the male; it is the second organism, because it is a plant, a hybrid fixed in the form of femininity.[27]

From this reordering of gender categories, Oken drew a number of startling conclusions. One was to return to what Laqueur called the one-sex model of gender—namely, the idea that the female sexual organs were but imperfect copies of the male's, a position Laqueur considered virtually outmoded by the time Oken was writing. Oken's argument is convoluted but goes something like the following: since the female sex, as negation, is a hybrid of the male and female, and since all living matter is a form of male animalcules, then there is no such thing as a pure female sexual organ. For that reason, "according to their organization, the female genitalia [are] based on their nature demand the masculine form."[28] The female sexual organs are nothing else, he claimed, than disorganized male organs.[29]

Oken's desire to relate everything to the male required him to provide an explanation for the birth of females in animals. In direct contradiction to the vitalists, Oken's central operating assumption, as analyzed by Vienne in this volume, concerning impregnation was that the female contributed nothing substantial to the process except to provide the place where the instantaneous formation occurred. This happened when the semen attached itself physically to the womb. The female's sole function in generation was formative and nutritive with the womb as the container where the polarization process occurred and shaped the form of the new creature. Oken, for the first time at a loss for an explanation, could not say why this was so. Lamely, he answered, "Asking why . . . is similar to asking why the stomach digests and the liver secretes gall."[30] The womb's nutritive function was, he assumed, clear. But if the female contributed nothing substantial to creation, how does one explain the birth of females?

Male birth was, Oken believed, easy to explain. The male sex is simple, independent, and only produces itself. However, in the animal, the male must reproduce himself within a female, and the female was a hybrid, a combination of male and female. This meant that the female had a tendency to transmute the male animalcules into a female form, because the formative process took place in the womb. Therefore, the production of male or female was dependent on the power of the female principle. "The more feminine the female, the more likely, everything else being equal, she would be to bring a girl into the world; the more masculine the female, the more likely a boy." Oken added, the definition of femininity did not depend on physical characteristics, but rather on powers and also on the strength of the male with whom she mates. If both principles were of equal power, then the woman would produce a male because "her inborn masculinity would tip the balance."[31] Once again we see the shift in conceptualization that has taken place between Enlightened vitalism and romantic *Naturphilosophie*. The vitalists considered all humans to be hybrids—mixtures of male and female powers—and strove for harmonizing both in a higher unity. The *Naturphilosophen* located hybridity in the female and considered it a fault, a falling away from a perfect form: pure (male) was good, mixed (female) was imperfect.

Die Zeugung presents us with a complicated attempt to rewrite the story of generation in order to recast gender relations. Gender difference was a universal law of nature, its traces evident in the first appearance of matter.[32] Oken set up stark dichotomies differentiating males from females in all moments of creation, establishing a set of hierarchies based on the original duality between polyps and plants. Thus, for example, he placed worms and birds under the active polyp category; insects and fish were placed under the passive category of plants. He even set up hierarchies within the living system based on the male/female separation, ranking them either as male (e.g., the respiration system) or female (e.g., the lymph system). The metaphoric language Oken used to describe each of these dyads reinforced this strict hierarchy. Males were active, creative, self-sufficient, independent; females passive, imitative, dependent. There was no possibility of an overlap between these dyads. The oxymoronic formulations of the Enlightenment vitalists disappeared and were condemned as major errors in correct scientific thinking, breaches in logic. Oken's major theme—that females can only live in and through males, which ran like a Leitmotif throughout the whole work—testified to the major shift in gender conceptions that separated him from Humboldt, Blumenbach, Buffon, and the Enlightenment vitalists.

THE "FEMALE AS SUCH": CARL GUSTAV CARUS

Oken's vision of gender, though individual (and at times extremely weird), was not unique. Its appeal can be seen in the work of later *Naturphilosophen*, such as Carl Gustav Carus, whose career extended well into the 1860s. Carus was educated as a physician and became a leading gynecologist in Dresden in the second decade of the nineteenth century. He trained scores of young men in this discipline, who spread out over Germany, degree and forceps in hand, empowered to treat women and to discourse about them.[33] In a late work published in 1861, Carus remarked that, though he had been inspired by the new movement initiated by Schelling and Oken during the first decade of the century, he had always maintained a degree of skepticism concerning the excesses of *Naturphilosophie*.[34] If that were the case, then Carus's narrative of gender, presented in a textbook on gynecology written in 1820, and reprinted many times afterward, offers a startling confirmation that Oken's position, despite its excesses, was highly appealing and readily accepted even by the more skeptical *Naturphilosophen*. Carus's textbook, bearing the title, *Lehrbuch der Gynäkologie*, purported to investigate "the teachings concerning the specifics of the female body, its construction, its life, its sicknesses and its respective dietetic and medical treatments."[35] But before Carus addressed these specific medical questions, he presented an introductory section dealing with the female as such. Its overriding imperative was to confirm and even to strengthen the difference between the male and female adumbrated in Oken's work.

Carus began by a discussion of the female's physical characteristics. The female body in comparison to the male's, he claimed, presented a picture focused on assimilation and reproduction. But more than that, it revealed a tendency toward a type not fully developed, which can be seen as being similar to a child's body. This linkage of the female with the child was supplemented and strengthened by relating the female's body to "lower species," or "incomplete organizations." The comparison Carus preferred here was the fish, which was for Oken, one of the prime embodiments of the female principle. And to make his point even clearer, Carus also equated the female to the plant, which he also considered radically opposed to the animal. The plant-female was directed to formation, maintenance, and metabolism, the animal-male to unity and freedom.[36] Throughout the whole female body, ranging from the rump to the brain, all Carus could detect, when compared to the male, was imperfection, childlike tendencies, and analogies to "less developed" species.

This excursion into the physical characteristics of the female, however, served as a prelude to the discussion of the female's psyche. For Carus, as for all of the *Naturphilosophen*, the body was not considered the determinant of the psyche, but rather the vessel of spiritual principles. The psyche was both "the true mirror of the physical, but even more the ideal side of the organism itself."[37] Thus, beginning from what Carus thought was observable nature—the shape of the female body, its vegetative nature, its similarity to the child and to less-developed species—he then applied these same characteristics to the psyche, which he argued consisted of three parts: the *Gemüth*, which he defined as the capacity to feel and experience; the *Geist*, the capacity to reflect and reason; and the will. The results read as though he were almost satirizing what later became the traditional gender stereotype of the late nineteenth-century *bürgerlich* German woman. The tragedy is that this was not satire.

On all three levels of the psyche, Carus considered the female far less capable and active than the male. Though the female possessed greater sensibility, more lively feelings, and a more active fantasy than the male, these features could never be developed to serve as a counterbalance to male superiority; they were always verging on the brink of either overexageration or inactivity. Thus, though the female was, Carus claimed, more sensitive to sounds, she lacked the ability to distinguish carefully between them, was incapable of acquiring a total tonal impression and hence possessed very little true musical ability. The same was true about the female's overly active fantasy and sensitivity; they got in the way of allowing her to produce great and elevated works of poetry and art. She had fantasy and sensitivity, but not the power of "creative fantasy," nor "deep feeling": everything was superficial. Since the male, in Carus's view, embodied the rational principle, the female was congenitally incapable of making any contributions to the sciences and the humanities. Being vegetative, the female also lacked the ability to act quickly, decisively, and arbitrarily; she was not able to combine all of the powers of body and spirit. The only tasks Carus believed the female better able to carry out were small ones, not requiring boldness and power, but patience and quiet, such as taking care of children, the sick, newly delivered children and their mothers.

The female's need to attach herself to a stronger person to compensate for her own weakness and lack of energy was translated in the sexual sphere, Carus argued, to a binding love for her husband and an even stronger love for her children (*Mutterliebe*)—feelings that already manifested themselves in the way little girls played with dolls. This desire also explained the explosions of hate and revenge of which she was capable if someone threatened her child or husband or if she was betrayed in love. The desire to please finally lay at the heart of the female's fascination with housekeeping and her mania for ornaments, attire, and fashion (*Putzsucht*).[38]

In the last section of this general discussion of woman's nature, Carus asked whether sexual characteristics were inborn, a matter of education, or perhaps stages in the development of the human from the embryo through childhood to adulthood. Simply said, he asked, were women born feminine and men born masculine? Carus acknowledged that some theorists, arguing for the unity of the human race, had posited the theory that the original embryo was of one sex and sexual differentiation occurred over time during epigenesis; some even argued that the original sex was female and hence it should serve as the norm against which men should be measured. True to his *naturphilosophic* assumptions, Carus denied all such theories, for generation was nothing but the appearance of already formed dualities.[39] Generation did entail the constant transformation of matter; there was continual emergence, generation, and disappearance, but never "a new creation, which is unthinkable."[40]

For Carus and the other *Naturphilosophen*, gender illustrated the hierarchies that constituted nature, hierarchies that he, as well as Hegel, Stone maintains, had linked to an emerging racial discourse. Though everything was in movement, nothing new was created. Everything pulsated within its own circumscribed sphere or circle. Each circle formed part of the infinitely graduated but continuous chain of being. All differences were analogues of the original difference, all reducible to a one-dimensional diagram charting abstract relations. The "realities" that the Enlightenment vitalists attempted to harmonize—time and space, matter and force, male and female, fact and imagination—became expressions of the one, the *Monas*. This identity was concentrated in the human being, who was the most perfect expression of the duality in the *trias*. The human was the "highest of the animals in whom everything that is noble and capable found distributed in the rest of living things is united; in him all the seeds and fruits, all matter and forms of the earth and heavens, the *avulsio ätheris*, flow together as in a focus."[41]

It has often been said that the philosophy of the Enlightenment had raised the human being to a dominant position over the rest of the living world. There may be something to such a critique, but in comparison to the *Naturphilosophen*, the Enlightenment's claims were modest indeed. For the *Naturphilosophen*, the ultimate focal point, where the noblest and best of human nature were joined was the white, European male. His reason, imagination, and willpower enabled him to pierce the veil of nature, and thus using his self-reflective consciousness, recapture its very origins and return to the womb from which he had sprung. This celebration of the male established the grounds for a type of instrumental reason far more dangerous than anything proposed by Enlightenment thinkers. In its very audacity, its elevation of theoretical reason over epistemological modesty, its denial of active individual agency, and its static and hierarchical worldview, *Naturphilosophie* was indeed a major rejection of the Enlightenment, but not the Enlightenment portrayed by many postmodernists. This is the final irony. For a careful look at the late Enlightenment might reveal a way of thinking and doing that is much more sympathetic to postmodernism than romanticism. The *Naturphilosophen* search for a new universal *mathesis*, in a world seemingly filled with chaos, found its clearest expressions in the definitions of gender (and later race) they had forged. What is even more striking is that these definitions outlived romantic natural philosophy itself, becoming inscribed in the new biology of the nineteenth century.

NOTES

1. Thomas Laqueur, *Making Sex: Body and Gender from the Greeks to Freud* (Cambridge, MA: Harvard University Press, 1990); Claudia Honegger, *Die Ordnung der Geschlechter: Die Wissenschaften vom Menschen und das Weib, 1750–1850* (Frankfurt/Main and New York: Campus, 1991).
2. Peter Hanns Reill, *Vitalizing Nature in the Enlightenment* (Berkeley and London: University of California Press, 2005).
3. Laqueur, *Making Sex*, 150.
4. Ibid., 154.
5. Anne-Charlott Trepp, *Sanfte Männlichkeit und selbstständige Weiblichkeit: Frauen und Männer im Hamburger Bürgertum zwischen 1770 und 1840* (Göttingen: Vandenhoeck und Ruprecht, 1996).
6. Wilhelm von Humboldt, *Gesammelte Schriften*, ed. Albert Leitzmann, 15 vols. (Berlin: B. Behr's Verlag, 1903), vol. 1:311.
7. Ibid., 314.
8. Ibid., 328–329.
9. Ibid., 329.
10. Ibid.
11. Ibid., 321.
12. Ibid. In this reverie, Humboldt indulged in sadistic dreams of control and enslavement. Yet, it is interesting that the object of this fantasy was a physically strong women and not the passive, slave girl model common to such fantasies.
13. Ibid., 322.
14. Ibid., vol. 15:451–460.
15. Ibid., vol. 1:328.

16. Ibid., 333, 334.
17. This is a point made clearly by Alison Stone in her chapter in this volume on "Sexual Polarity in Schelling and Hegel."
18. Friedrich Wilhelm Schelling, *Erster Entwurf eines Systems der Naturphilosophie: Zum Behuf seiner Vorlesungen* (Jena and Leipzig: Christian Ernst Gabler, 1799), 43.
19. See Stone, "Sexual Polarity," in this volume.
20. Oken, *Die Zeugung* (Würzburg: Goebhardt, 1805), 110.
21. Ibid., 23.
22. Ibid., 51.
23. Ibid., 216.
24. Ibid., 112–113.
25. Ibid., 124.
26. Ibid., 13.
27. Ibid., 133.
28. Ibid.
29. Ibid., 181.
30. Ibid., 104.
31. Ibid., 140–142.
32. Ibid., 189.
33. Carus, born in 1789, became professor for "Frauenkunde" in Dresden in 1814 and also director of its lying-in hospital. He held these positions until 1827 when he was appointed physician to the king. He died in 1869.
34. Carl Gustav Carus, *Natur und Idee, oder das Werdende und sein Gesetz: Eine philosophische Grundlage für die specielle Naturwissenchaft* (Vienna: W. Braunmiller, 1861), v.
35. Carl Gustav Carus, *Lehrbuch der Gynäkologie oder systematischen Darstellung der Lehren von Erkenntniß und Behandlung eigenthümlicher gesunder und krankhafter Zustände, sowohl der nicht schwangern, schwangern und gebärenden Frauen, als der Wöchnerinnen und neugebornen Kinder* (Leipzig: G. Fleischer, 1820), 4.
36. Carus, *Lehrbuch*, 41.
37. Ibid., 46.
38. Ibid., 40–48.
39. Ibid., 48–49.
40. Ibid., 41.
41. Oken, *Die Zeugung*, 1.

4

ZEUGUNG/FORTPFLANZUNG

Distinctions of Medium in the Discourse on Generation around 1800

JOCELYN HOLLAND

By now it has become commonplace to say that early German Romanticism, with its penchant for unusual couplings, was equally informed by the arts and the emerging life sciences. The same holds true for one of romanticism's most prominent areas of experimental thinking: the discourse of procreation, or *Zeugung*. Due in part to a reluctance to accept conventional divisions between the organic and the inorganic, or between the material and the abstract, and in part to an experimental spirit, which deliberately sought productive juxtapositions, the language of procreation appears in the most unexpected places. To this end the German Romantics, and in particular Friedrich von Hardenberg (Novalis), Friedrich Schlegel, and Johann Wilhelm Ritter, draw from sources as diverse as Immanuel Kant's discussion of the organism in the *Critique of the Power of Judgment*, Johann Gottlieb Fichte's philosophy of the self-positing subject, the reproductive theories of Johann Friedrich Blumenbach and Carl Friedrich von Kielmeyer, Antoine Lavoisier's writings on chemistry, and the mystical writings of Jacob Böhme, to name just a few.[1] *Zeugung*, as it was appropriated and defined by Romanticism through a rejection of stable hierarchies and categories, was not, however, the only concept related to procreation under discussion at the end of the eighteenth century.

Part of the challenge in understanding the complexity of procreative thinking around 1800 lies precisely in the fact that there are two concepts in use that share an uneasy kinship. In addition to *Zeugung*, *Fortpflanzung* is also alleged to perform the conceptual labor of procreation. Joined at times to the point of indistinguishability, these two concepts are nonetheless indebted to different points of departure and serve different purposes. *Zeugung* originates in the context of animal generation and is defined by what one early nineteenth-century dictionary calls "the activity wherein one brings forth one's own."[2] Indebted to a logic of self-similarity, which lends itself to the autopoietic fantasies of romantic literature, *Zeugung* tends to emphasize the production of a new individual in terms of its kinship with the parent. *Fortpflanzung*, in turn, was initially linked to the context of plant reproduction before evolving to include the propagation of other things,[3] including the various categories of procreating individuals, be it a genus, a species, or—from a theological perspective—those stained by original sin. If *Zeugung* produces an individual (or, more figuratively, an idea, or a work of art), which is somehow indebted to its parent or creator, the emphasis of *Fortpflanzung* is, through the incorporation of singular acts of *Zeugung*, to reproduce part-whole relationships on a larger scale by maintaining the collective, which includes the individual.

Both of these concepts of generation have an established history in the German language. To take the Luther Bible as an example, one could compare the chains of genealogical production whereby a parent *zeugte* a child[4] to the *Fortpflanzen* of a sin, a family and the gospel, sometimes with direct reference to the generation of plants.[5] One corollary to this distinction is that *Zeugung*, as the production of the individual, tends to be figured within the moment, collapsed in time and space, whereas *Fortpflanzung* is defined by duration and expansion. Another claim examined in the following discussion is that *Fortpflanzung* lends itself to narratives of production and generation that are less conditioned by biological concepts (like *conception* and *birth*, as one finds in the poetics of *Zeugung*) than they are by the integration of discrete moments of individual production within a constructed continuum, which both defines and is defined by the individuals. How the continuum of propagation is defined and what it is called is *context-specific*, but it exists as the virtual counterpart to the physical production of the individual. Friedrich Schelling, in his *Erster Entwurf eines Systems der Naturphilosophie* (First outline of a system of the philosophy of nature, 1799), notes that "the blossom wilts" and "the metamorphosed insect dies away, as soon as the species is secured," such that "the individual seems here almost to serve merely as a medium."[6] According to Schelling's observation, the concepts of *Zeugung* and *Fortpflanzung* could be said to enjoy a peculiar relationship. Whereas *Zeugung* is directly accessible through the physical medium of the organism or the materiality of objects in the world, *Fortpflanzung* is indirectly accessible through a different kind of "medium," which only becomes visible—that is, is granted a form—through the appearance and disappearance of physical bodies.[7] As fleeting as Schelling's observation may be in the broader concept of his work, it helps open up an inquiry into the relation between these two concepts around 1800.

The goal of this chapter is essentially to expose the intricate relationship between the concepts of *Fortpflanzung* and *Zeugung*.[8] It begins with a brief overview of the role played by *Fortpflanzung* and *Zeugung* in physiological and philosophical thought around 1800 with reference to Goethe, Herder, and Fichte. The second part of the chapter then focuses on examples of the discursive "propagation" of *Fortpflanzung* from the same time period. Taking two case studies drawn from chemistry and acoustics found in the work of Johann Wilhelm Ritter, it examines the degree to which *Fortpflanzung*, when transplanted into inorganic regimes for the purpose of scientific investigations, still functions in a manner comparable to its role in the organic-philosophical context from which it emerged. As we see, Schelling's notion of the individual as a fleeting medium for the species is revisited in surprising ways. The concluding section then offers a brief footnote to the discourse of *Fortpflanzung* in the romantic context to show how key elements of the discussion around 1800 resurface unexpectedly in the biology of Karl Ernst von Baer.

FORTPFLANZUNG: MODES OF CONTINUITY IN GOETHE, HERDER, AND FICHTE

The distinction between *Zeugung* and *Fortpflanzung* was a topic of theoretical reflection at the end of the eighteenth and the beginning of the nineteenth centuries within the domain of organic reproduction, including plants and animals. When Johann Wolfgang von Goethe wrote the *Versuch die Metamorphose der Pflanzen zu erklären* (Attempt to explain the metamorphosis of plants) in 1790, he needed to distinguish between different phases of the plant's development: its growth (which he defined in terms of a single, mutable organ or *leaf*) and its reproduction. Goethe's essay navigates this problem by distinguishing between "growth" (*Wachsthum*) and "procreation" (*Zeugung*)[9] while joining these concepts to a more general notion of propagation. For example, he describes the actual moment of procreation in terms of a "*Fortpflanzung* through two sexes," one that occurs after the leaf in its various metamorphoses has ascended a "spiritual ladder" to the "pinnacle of nature."[10] He also uses the temporal categories of successivity and simultaneity to establish a conceptual hierarchy in which *Fortpflanzung* is the generic term:

> If we observe the plant insofar as it externalizes its life force, then we see this occur in two ways, first through *growth*, in that it brings forth stem and leaves, and then through the *Fortpflanzung*, which is completed in the building of blossoms and fruit. If we inspect growth more closely, we see that, as the plant reproduces (*fortsetzt*) itself from node to node and from leaf to leaf, as it vegetates, a *Fortpflanzung* may also be said to take place that distinguishes itself from the *Fortpflanzung* through the flower and fruit, which happens *all at once*, because it is *successive*, and that it shows itself in a sequence of individual developments.[11]

Goethe's theory of metamorphosis is concerned with a concept of *Fortpflanzung* in which the individual plant serves as a model for others but not explicitly for a species as a whole. At the same time, it includes several important elements composing a simple pattern of *Fortpflanzung*, which will be elaborated in further examples—namely: that it is indebted to a diachronic joining, which includes individual moments of unity, and that *Fortpflanzung* for that reason can be said to be both equal to and greater than *Zeugung*. Through its understanding of empirically observable physical growth in conjunction with a "spiritual"—one could say *virtual*—ladder, Goethe's text indicates how *Fortpflanzung* can also be construed as a medial concept. It is the diachronous coupling of elements, which validates the metaphor of the ladder in the first place. The connection between *Zeugung* and *Fortpflanzung* can therefore be thought of in various ways, each of which expresses a relationship between part and whole, whether in terms of time (*Zeugung* as one empirical stage of development in the *Fortpflanzung* of the organism), as a conceptual distinction (which permits *Fortpflanzung* to be greater than or equal to *Zeugung*), or in terms of the relation between a medium and its elements, where the spiritual or virtual dimension of *Fortpflanzung* is only revealed through empirical observation. What is more, each of these distinctions can be lifted out of its original setting of plant metamorphosis and reconsidered in an anthropomorphic context. For Goethe, the two share a natural affinity. His elegy on the metamorphosis of plants from 1798, where the progress of a human pair echoes the plant's climb up a "spiritual ladder," and where so much depends on the narrator's ability to convey a kind of thinking that moves from the empirical to the intuitive, can attest as much.[12]

The same connection between humans and plants is also prevalent in the work of Johann Gottfried Herder. His treatise, *Ideen zur Philosophie der Geschichte der Menschheit* (Ideas for a philosophy of the history of mankind, 1784/91), contains a chapter devoted to the relation of the plant realm to human history in which he begins by identifying common features between humans and plants that then give way to comparable phases of change over time. Throughout the chapter, individual processes are more directly implicated in a narrative for the species than was the case with Goethe.[13] The overall comparison maintains its predictive potential by subjecting initial forms to a series of parallel transformations without trying to cling to morphological similarities.

Herder's lengthy comparison changes dramatically, however, when he introduces *Fortpflanzung* into the discussion with its two inflections of a procreative act and a more general process of propagation. Like Goethe, Herder uses *Fortpflanzung* in each of these senses, not always clearly distinguishing it from *Zeugung*. Drawing from a vocabulary evocative of Charles Linné, he writes,

> [T]he blossom, we know, is the time of love for the plants. The sepal is the bed, the corona its curtains, the other parts of the flower are instruments of *Fortpflanzung* which nature has displayed openly on these innocent creatures and decorated with all glory.[14]

Through a static description, the flower acquires contour. Its space becomes more complex and acquires greater plasticity, attracting the observer's gaze to the center of the plant's reproductive activity before coyly diverting it. When Herder's narrative transitions from the elaborate description of propagation's "tools" to their function, simple declarative statements give way to a subjunctive: "[Nature's] great purpose should be reached." This purpose is then qualified as "*propagation, maintenance of the species*" (Fortpflanzung, Erhaltung der Geschlechter).[15] The rest of the chapter leaves little doubt that such a goal is endlessly attained—the subjunctive "should" simply acknowledges that propagation is the hinge on which the survival of the species turns. Ultimately, nature produces a multitude of seeds, far more than can survive in any single generation. The life of the species comes at the expense of some of the individuals (this is *Fortpflanzung* as *Zeugung*, successful or not), such that propagation is paired with death. On the other hand, in this passage and elsewhere in Herder's text, the emphasis is on *Fortpflanzung* as *Erhaltung*, with its register of continuity, perpetuation and sustainment.[16]

More overtly even than Goethe's "ladder," Herder's equation of propagation with "purpose" (*Zweck*) underscores its inherent teleology. That propagation is the means toward some end is also conveyed by the fact that *Fortpflanzung* (and *Erhaltung*) are usually located in a genitive construction—one speaks about the propagation *of* something *through* something to satisfy a purpose of nature. Hence, it comes as no surprise that *Fortpflanzung* also figures prominently in the work of those philosophers whose systems incorporated the perceived teleology of nature into their understanding of moral law. Such is the case with Fichte, where *Fortpflanzung* appears both in physical accounts of generation and in the context of a moral law, which enjoins us not to act against nature.[17] Fichte's writings on moral law and natural right unite the virtual and physical components of propagation, as can be shown by comparing passages from *System der Sittenlehre* (The system of the doctrine of morals, 1798) and the *Grundlage des Naturrechts* (Foundations of natural right, 1796).

In the former text, Fichte describes how the relationship of husband and wife is grounded in the natural disposition toward the propagation of the species (*Fortpflanzung des Geschlechts*). The desire to procreate is a "natural drive," which, he argues, only exists for itself and its own satisfaction. From the perspective of human nature, it is a purpose (*Zweck*) and from the perspective of nature more generally, it is a means (*Mittel*) to ensure the continuation of the species.[18] The *Foundations of Natural Right* contains similar language. A species propagating itself cannot be thought of as an "eternal becoming" of forms. Were it always in transition, there would be no possible concept of being; no fixed points against which to discern change at all. To continue, the species must have another organic existence, at which point Fichte introduces the idea of the propagating individual:

> If nature were to be possible, the species had to have some organic existence other than its existence as a species; but it also had to exist as a species, so as to be able to propagate itself. In order for this to be possible, the species-forming power had to be divided up and split into two perfectly matching halves, as it were, whose union alone would constitute a self-propagating whole. In being divided this way, the species-forming power forms only the individual. . . . It is only the individuals (in their union and capacity to be brought into union) that exist, and only they that form the species.[19]

This passage from the *Foundations of Natural Right* is remarkable for capturing the elusive nature of *Fortpflanzung*'s form/medium relation. There is a double, related emphasis on "existence": one on the existence of the concept *species* (i.e., the circumstances under which the concept of a particular species can be said to exist) and the other on the elements of organic existence that validate and occasion the concept. At the same time, there is another aspect to this passage, whereby the species precedes and creates the individuals, as if species were not an anthropocentric concept at all. There is thus a tension in this claim between the creation of the category "species" and its function: the pair of procreating individuals both *form* the species and *are* the species. They create it and it defines them reciprocally.

Taking a step back, when we look at the ways in which Goethe, Herder, and Fichte navigate the relationship between *Zeugung* and *Fortpflanzung*, it is possible to see certain tendencies emerging as well as patent differences. Goethe, though less engaged with the species concept in his writings on plant metamorphosis than Herder and Fichte, articulates a relationship of part and whole compatible with the other two thinkers, where the individual moments of *Zeugung* are more than encompassed by the whole of what *Fortpflanzung* can signify. Herder, who draws extensively from the discussion of plant metamorphosis in his discussion of human life, is more concerned with the bigger picture of individuals and species, sharing Schelling's idea of the latter, which exists at the expense of the former. The same can be said of Fichte. In short, despite varying concern for individuals and species and for moments of procreation as embedded in the propagation of a whole, and despite greater or lesser emphasis on a teleological framework, it is still possible to observe the presence of a mutual set of concerns, articulated in the language of organic growth (and decay). In the work of Johann Wilhelm Ritter we see how these same ideas are tested and developed in the inorganic realm.

FORTPFLANZUNG'S PROPAGATION: JOHANN WILHEM RITTER ON ACOUSTICS AND CHEMISTRY

One of the most striking differences between *Zeugung* and *Fortpflanzung*, as between procreation and propagation, is the latter's greater success in expanding into discourses far beyond the immediate context of plants and animal generation. Although for romantic poetics it was legitimate to speak of the procreation (*Zeugung*) of inorganic materials, it was not common practice at the time to speak of the "procreation" of sound, light, and electricity; nor was it common to consider their propagation (*Fortpflanzung*) as anything more than a convenient metaphor of extension and continuity. The work of Johann Wilhelm Ritter comprises an unusual episode in the history of these concepts, however, because of his various attempts, particularly in the fields of acoustics and chemistry, to visualize a kind of "real" *Fortpflanzung* in inorganic contexts that would be more than a figurative propagation. At the same time, he also preserves the distinction between the physical medium of propagation (i.e., the elements which compose it) and the concepts or "forms," which are constructed during the process. The remainder of this section focuses on Ritter's attempts to preserve a rigorous notion of *Fortpflanzung*, even in nonorganic contexts with reference to two case studies drawn from his *Fragmente aus dem Nachlasse eines jungen Physikers* (Fragments from the estate of a young physicist, 1810) and his essay on the history of chemistry.[20]

Before testing out Ritter's notion of a real propagation in the inorganic realm, it is helpful to first take a look at how this idea would function in an organic context. We can take as an example one of Ritter's fragments that underscores the common distinction between a physical *Zeugung* and a virtual *Fortpflanzung* to an even greater degree when he envisions a different kind of organic economy—one defined by accumulation instead of exchange:

> Everything which exists maintains itself [*erhält sich*] organically. Every rock emerges new in every moment, produces itself onward into the infinite. Only the parents of the child die at once, again and again, and thus one does not see the individual increase. If one can undo the annihilation, then the new attaches itself to the remaining old, and now real *Fortpflanzung*, accumulation, takes place.[21]

The logic of *Zeugung* is one of equilibrium: for every birth, a death. To defy this logic, as Ritter's fragment speculates, would permit an accumulation to take place through the coexistence of old and new. Ritter's suggestion that one would finally have a real *Fortpflanzung* in this particular scenario serves to emphasize the idea of a virtual one; that is, apart from such fantasies of perpetual life, *Fortpflanzung* is a concept that exists only as an idea through (but also at the expense of) the individual. When he imagines an accumulation of individuals in space who would normally be separated by their intervals in time, Ritter does more than challenge the distinction between a figurative and a literal *Fortpflanzung*, however. He also underscores the anticipatory dimension of the "form" of *Fortpflanzung* in that its temporality extends beyond our immediate faculties of perception and experience in a way which is different than with *Zeugung*. The following examples, like those we have seen so far, take up the question of how individual elements participate in a medium. They also attempt to account for the dual nature of *Fortpflanzung* as the production of the individual and of the category within which it exists.

Ritter's Acoustics

Experiments in the propagation of sound, light, heat, and electricity had been conducted with increasing frequency since the end of the seventeenth century. Common to each of these diverse experimental environments is a concern for the medium (such as air, water, or sand) in which the propagation takes place. The relative opacity or transparency of the medium directly affected the degree to which the propagated entity is facilitated or hindered in its movement. Acoustical experiments studied the propagation of sound (*Schallfortpflanzung*) with regard to the relative density of the surrounding air as well as other substances. In particular, the work of Ernst Chladni was influential for rethinking acoustical phenomena in terms of visible traces. The so-called Chladni tone-figures (*Klangfiguren*) are the products of sound waves caused by reverberating strings brought in proximity to sand. Both Ritter and other romantic thinkers saw in the Chladni tone-figures a kind of natural hieroglyphics waiting to be deciphered. The appendix to Ritter's fragment project, which concerns itself exclusively with acoustical phenomena, is paradigmatic for the romantic reception of Chladni's work. With reference to both Chladni and Hans-Christian Oersted, a friend of Ritter's who also worked in this field, the appendix synthesizes the experiments of these two scientists with the speculative thinking of Johann Herder. On the basis of two analogies, Ritter comments on the intimate connection between sound, writing, and organic life. The first is that every tone has its "letter," or unique trace, as a vibrating string held in close proximity to sand or water will illustrate. To justify this connection, Ritter draws upon Herder's text, *Die Älteste Urkunde des Menschengeschlechts* (The oldest document of the human race, 1774/76) and the *Abhandlung über den Ursprung der Sprache* (Treatise on the origins of language, 1770).[22] The second analogy is that every tone is its own being, "an entire organism of oscillation, figure, and shape, just like every organic-living [thing]."[23] It is on the basis of this second analogy that the connection between biological and acoustical *Fortpflanzung* unfolds. If tones are "organisms" that, like living beings, have a duration in time and interact with others of their kind, then, analogous to humankind, they will participate in a larger collective. The confluence of sound, letter, and being thus allows Ritter to imagine a "genealogy" of tone as an acoustical human history: "Entire histories . . . of peoples, indeed the entire human history, must allow itself to be expressed musically."[24]

We can observe that the range of these acoustical speculations, which lead from the production of tone to writing and an imagined genealogy, imply a displacement of medium from inert material to the human body. With the notion that tones, as organisms, can articulate "entire histories," we have taken a step away from simple patterns in the sand. Bettina Menke has discussed Ritter's idea in terms of an aesthetics of listening and a semiotics of writing:

> If tones as mere motion—movement, which moves—first record and inscribe themselves in Chladni's "tone figures" as analogue [data], then hearing means the effect of the self-inscription of the oscillatory movement, which is the sound. To this corresponds Johann Wilhelm Ritter's suggestion: "Every tone is therefore immediately accompanied by its letter", so that we "really only *hear* writing" and that we "read, when we hear"—as Ritter adds. This displaces the concepts of writing and reading, however, as it were beyond the state of being relayed by signs which *comprises* them, toward [1] a self-inscription which would be the *Fortpflanzung* and reproduction of motion, and toward [2] an unconscious "perception" of an inscription which has no subject, an operation regarding distinctions which no longer have a subject but which make bodies.[25]

Menke's primary concern in this passage is to describe a new kind of writing, which is neither the work of a self-aware subject nor the "natural writing" of the "book of nature," that refers back to a divine hand. According to Menke, to claim that we "read when we hear" is to accept that we perceive a kind of text separate from any act of writing—a text which is simply recorded by us without our active perception of it and transmitted to us without requiring an agent. Within this constellation of acoustic production and reception, the producing agent is no longer intact, yet the biological metaphors of propagation are still present (i.e., through reference to human history, and a genealogy of tone). As a result, the concept of *Fortpflanzung* necessarily takes on new meaning. It becomes pure operation and motion, and the resulting (receiving) subject is merely the medium of inscription. In other words, the scientific discourse of acoustical experiments allows us to imagine an unusual decoupling of the two concepts—a pure *Fortpflanzung* without

Zeugung—where the organism in question is only medium and duration. The degree to which Ritter was willing to entertain this possibility is evident in the following passage:

> In the *tone*, generally, where discontinuity in time *nonetheless* produces continuity, would be the first place to ask whether *different qualities which border one another merely in time* . . . could perhaps not also produce *processes* and processes of a *particular* kind among each other?—Already something similar lies in every *Fortpflanzung* of something, since namely here a time still really passes, until the process, etc., of *a* in *b*, of *b* in *c*, etc., has arrived.²⁶

The difference in the temporality of this metaphor of acoustical *Fortpflanzung* and, for example, the one in Goethe's discussion of *Fortpflanzung* and *Zeugung* in the life of the plant is striking. For Goethe, *Zeugung* was characterized by simultaneity, as the event of a moment, and the *Fortpflanzung* of the species was characterized by a constant sequentiality and duration. In Ritter's acoustical speculations, however, his fantasy of a real *Fortpflanzung* is actualized. The propagation of sound can overcome temporal lapses because tones generate their own unique processes—"discontinuity . . . produces continuity." As elements, the tones allow for a form to be propagated, which endures in time. In other words, the essence of *Zeugung*, as the production of one's own (invoked earlier in the phrase *seines Gleichen*, and here, *eigner Art*) has been reworked into the logic of propagation. Though grounded in acoustical phenomena, the tendency of this passage is to take the acoustical model of *Fortpflanzung* and generalize it: If one can observe this process in a series of tones, then, Ritter asks, why not in every *Fortpflanzung*?

Ritter's Chemistry

Ritter's comments about the *Fortpflanzung* of sound and light resonate in a peculiar way when he writes about the pile—an early form of the battery, which could conduct an electrical current and thus perform electrolysis on substances like water. Ritter describes the ability of the pile to separate water into its constituent elements of oxygen and hydrogen before having them recombine as water as a *weiterpflanzen*, where the substance propagated was water itself: "And what the [pile] produced was again nothing other than—*water*. To divide *this* in order to unite it again was its single preoccupation."²⁷

At the same time, there is an important distinction to observe between chemical and acoustic propagation in terms of medium: the propagation of sound and light occurs in the more or less transparent media, but in the case of the pile, which conducts an electrical charge, the thing that is propagated is also the medium itself. Ritter was fascinated with this aspect of the pile and by experiments conducted by chemists, such as Humphrey Davy, where acids and bases were thought to decompose on one pole of the pile and then recompose on another, crossing over in diluted form in between. Ritter even praises these chemical experiments for demonstrating a "true organic *Fortpflanzung*."[28] What would it mean for the processes of decomposition and recomposition on the pile to be understood in terms of organic propagation, as Ritter claims? His thinking proceeds under the assumption that, in mechanical production, a body reconstructs itself "in place," whereas in organic procreation a body disappears in one place and returns in another, such as through the production of one's offspring.[29] Ritter realizes that the idea of inorganic material obeying the laws of organic procreation (i.e., that mechanical propagation, or locomotion, where spaces are exchanged, should follow organic laws) will be met with skepticism. He assures his readers that he does not want to rehash alchemical theories. Rather, his purpose was to consider the possibility that, in its earlier state, inorganic matter might have undergone the same kind of production as organic matter and that remnants of this earlier state can still be observed in a few chemical processes. It is an idea he tests out in depth in the chemistry essay, thereby rewriting both the history of chemistry and the history of propagation itself.

In a lengthy footnote, Ritter raises the possibility of an inorganic *Zeugung* and *Fortpflanzung*, which would still operate in accordance with organic laws. In his somewhat convoluted prose, this is articulated as the wish to "extract as if thread the act of the 'eternal' reconstruction of everything ever existing," even if, he says, "only as a *consequence* of such a [reconstruction] it *has* this appearance of an existence."[30] In other words, intending to investigate inorganic material for the patterns of organic reproduction, he envisions an experimental situation, which would allow him to study the dual components of reproduction: the physical disappearance and reappearance of the body (the logic of *Zeugung*) and the virtual "thread," which symbolized "eternal reconstruction" (the logic of *Fortpflanzung*). He lists a series of examples that seem to approximate the organic model, including the chemical process of penetration (*Durchdringung*) and cementation. Ritter is looking for examples of processes where matter disappears in one place (not just through transposition, but where it really diminishes to a zero weight) only to reappear in another.

As his prime example, he returns to the transport of acids and bases from one end of the pile to the other. The key question here is: Do they disappear and reappear entirely or are they transported in reduced form? It is an important philosophical question for Ritter, one that he cannot answer entirely to his satisfaction. For the purposes of the present discussion, it is sufficient to note that his work comprises a special moment in the conceptual history of *Zeugung* and *Fortpflanzung* for at least two reasons. The first regards his willingness to take the propagation of *Fortpflanzung* into inorganic regimes as matter worthy of serious scientific study, such that propagation in the context of sound or chemical substances is not just to be understood figuratively. The other has to do with his experimental thinking on the concept of *Fortpflanzung* itself, which shows how it is precisely the act of transporting it outside of the organic context that leads to intriguing hypothetical possibilities, such as a "real" or pure propagation, which does not rely on procreation at all.

VIRTUALIZING PROPAGATION: KARL ERNST VON BAER

Propagation understood as an overcoming of ruptures in the endless cycles of procreation and death (i.e., propagation as the production of categories of procreating creatures) and the human as the one creature for whom procreation is not a matter of instinct but rather free will and therefore governed by moral law, which still conforms to the natural law of propagation, are ideas that appear in Goethe, Herder, and Fichte, and come together in the work of Karl Ernst von Baer, one of the preeminent biologists of the nineteenth century, to whom the discovery of the mammalian ovum is credited. If Ritter receives the credit for radical experimental thinking, which—based in a romantic context—transposes the logic of the organic realm into the inorganic, in von Baer, one can observe a reception of romantic thinking, which never strayed far from the bounds of mainstream science. Von Baer's 1834 lecture "Das allgemeine Gesetz der Natur in aller Entwicklung" ("The general law of nature in all development")[31] was conceived as a synthesis of the physiological theories of the day. It is of particular interest here as a document that bears echoes of a romantic engagement with propagation and procreation within the broader work of an (institutionally) established biologist, taking as his point of departure the inherent "self-destruction" (*Selbstzerstörung*) of life: every procreation (*Zeugung*) brings with it a "death sentence."[32] After citing the transience of the natural world, however, where the offspring of some species are not permitted to exist more than a single day, von Baer notes that there is something that persists over time. "An investigator of nature," he writes, "recognizes that we possess an innate necessity to believe in something eternal, even when we do not immediately perceive it."³³ What von Baer has in mind, however, is not necessarily (nor is not limited to) belief in God and eternity, which, he says, we are in any case not able to comprehend. Instead, the perspective we acquire by turning our gaze back to the natural products of the world is that,

> [T]he monads like the corporeal human are only transitory phenomena, which however during their fleeting existence, compelled by change, prepare the seeds for its [i.e., existence's, J.H.] renewal in other individuals, the passing realizations of a persisting thought, for through the procreations (*Zeugungen*) the same organization is still repeated . . . The *forms of organization*, those rows hanging together through procreation (*Zeugung*), appear as the persisting thoughts of creation; the individuals are passing representations of these thoughts.³⁴

Von Baer's choice of the word *rows* (*Reihen*) to describe the continuity of "forms of organization" simplifies the plurality of individuals within any given species into a single chain of repetition through procreation (an idea expressed even more succinctly in another passage, which refers to a primal procreation for each species followed up by its *Fortpflanzung* over time). To phrase it more pointedly, in the context of von Baer's earlier comments about how death is inscribed in every act of procreation, one could say that the mediality of such a row is defined by the ability to construct and sustain the thought of sameness (like Adelung's self-similarity of parent and child) despite ever-recurring ruptures and thus provide a counterpart to the "lasting thoughts of creation." This idea gains traction in seemingly disparate parts of von Baer's argument, such as in the case of maternal bodies, which die in the moment of giving birth, and the analogous notion of the continuity of youth and old age, even though "nothing is of the youth in the man, nothing of the child in the youth, other than a similarity of form and the memories of earlier experiences, which the self-consciousness gathers into the unity of the ego."³⁵

This example of (psycho-)physiological thinking from the 1830s is closer to the Fichtean example of *Fortpflanzung* in the moral context than it might appear at first glance. The only distinction of note between humankind and (other) animals in von Baer's lecture is the ability to control one's desire to reproduce, and the distinction between *Zeugung* and *Fortpflanzung* is particularly relevant here. According to von Baer, an animal is driven not to the single act of procreation but to *Fortpflanzung*: "The animal is therefore driven to *Fortpflanzung* through instinct."[36] Yet once creation had decreed humankind's gift of free will, "thus he also could exercise control over *Zeugung*."[37] The animal serves nature's purpose blindly, but the human is granted the ability to act through reason and serve nature deliberately.

CONCLUSION

We see in the work of Goethe and Herder, as well as the philosophers, such as Fichte and Schelling, an attempt to define *Fortpflanzung* as the perpetuation or maintenance (*Erhaltung*) of something through the production of forms that provide indirect evidence of the ineffable medium of propagation. These forms become apparent as a spiritual conduit (*Leiter*) and through the formation and lastingness of concepts such as *species*, which supersede individual breaks in the chain. It is the biologist Karl Ernst von Baer who summarizes this phenomenon most succinctly when he refers to propagation in terms of the production of a *thought*—an intellectual, cognitive activity that logically accompanies biological change. Ritter, for his part, conjures the fantasy of a pure *Fortpflanzung* without *Zeugung*, which envisions a propagation that would essentially complete the work of procreation without succumbing to its logic of ruptures—of endless births and deaths—to which our human existence is indebted. In this sense his work goes against the grain of predecessors, such as Goethe, for whom the concepts of *Zeugung* and *Fortpflanzung* are ultimately too tightly intertwined to be separated without undue manipulation, even though Ritter's thinking about procreation draws from Goethe's work on plant metamorphosis. Ritter's model of procreation imagines a scenario without interruption and without time in exchange for infinite accumulations. His visions of inorganic propagation that act procreatively occupy both the origin of procreation (to the degree that present examples in the field of chemistry are taken as evidence of past existence) and threaten to usurp the still sanctified space of the romantic subject, which requires yet further rethinking and redefinition once it is reduced to a continuum of propagated inscriptions, as we saw in the examples of *Fortpflanzung* in acoustical regimes. Without a doubt, Ritter's work occupies the periphery of romantic procreative thinking, but the work of more "mainstream" writers also testifies to the fact that the discourse of *Fortpflanzung* has a dynamics all its own that diverges from that of *Zeugung* in interesting ways. Its medium is not to be reduced to a physical substrate, as is the case with procreation understood as *Zeugung*. Even if the nuances and complexities surrounding the discussion of *Zeugung* and *Fortpflanzung* were otherwise not maintained by nineteenth-century biology, they constitute an attempt to think through a relationship that had been taken for granted, and whose reflection eventually comprised an important contribution to the romantic discourse on procreation.

NOTES

1. I discuss these connections in greater detail in *German Romanticism and Science: the Procreative Poetics of Goethe, Novalis, and Ritter* (New York: Routledge, 2009).
2. "die Handlung, da man seines Gleichen hervor bringet," in Johann Christoph Adelung, *Grammatisch-kritisches Wörterbuch der hochdeutschen Mundart* (Wien: B. Ph. Bauer, 1811), s.v. "Zeugung."
3. "vermehren, fortdauern machen, eigentlich von Pflanzen, in figürlichen Verstande auch von Thiere, ja von fast allen Dingen," in *Grammatisch-kritisches Wörterbuch*, s.v. "fortpflanzen."
4. For example, "Noah war 500 Jahre alt und zeugte Sem, Ham und Jafet." Gen. 5:32, *Die Bibel oder die Heilige Schrift des Alten und Neuen Testaments nach der Übersetzung Martin Luthers* (Stuttgart: Deutsche Bibelgesellschaft, 1979).
5. For example, Micha "welcher hernachmals, als er heranwuchs, vier Söhne zeugte und Saul's Geschlecht bis in die nachexilische Zeit fortpflanzte," 2 Samuel, 9:12, *Die Bibel*; and in the *Tischreden*, Luther interprets John 14:12 as, "Ich bin nur ein Senfkörnlin, ihr aber werdet sein wie die Weinstöcke und Aeste oder Zweige, auf welchen die Vogel werden nisten" to mean that the followers of Christ will preach "auf den Dächern und offentlich in der ganzen Welt . . . und das Evangelium allen Menschen verkündigen und fortpflanzen," in *Dr. Martin Luther's vermischte deutsche Schriften II: Tischreden* ed. Johann Conrad Irmischer (Frankfurt/Erlangen: Heyder and Zimmer, 1854), vol. 1:94–95.
6. "Die Blüthe verwelkt, das verwandelte Insekt stirbt dahin, sobald die Gattung gesichert ist. Das Individuum scheint hier fast bloß als Medium zu dienen, durch welches jene organische Erschütterung, nur als Leiter, woran die bildende Kraft (der Lebensfunke) sich fortpflanzt." Friedrich Wilhelm Joseph Schelling, *Erster Entwurf eines Systems der Naturphilosophie*, in *Werke: Historisch-kritische Ausgabe*, ed. Wilhelm G. Jacobs and Paul Ziche (Stuttgart: Frommann-Holzboog, 2001), vol. I, 7:n106.

7. There are various theoretical models available, and the use of *medium* and *form* in this chapter is shaped by Luhmann's discussion of these terms (which, in turn, is based in part on the work of Fritz Heider). For Luhmann, the medium is characterized by a loose coupling of elements. These can be considered independently of each other, but they also accommodate closer couplings, which result in the manifestation of forms. See Niklas Luhmann, *Die Gesellschaft der Gesellschaft* (Frankfurt: Suhrkamp, 1998), 190. In the context of procreation, one could say that the reproduction of individuals over time (as elements) is required for the perception of categories such as the genus, the species, and the like (as linguistic forms).
8. Those who are interested in the more general notion of reproduction should refer to Susanne Lettow's chapter in this volume, which also explores the temporal dynamics of reproductive theory. Readers who wish for more insight into the constructions of gender categories, which accompany romantic theories of generation, particularly in the work of Wilhelm von Humboldt, Lorenz Oken, and Gustav Carus, are advised to turn to Peter Hanns Reill's contribution.
9. Johann Wolfgang von Goethe, *Zur Naturwissenschaft überhaupt, besonders zur Morphologie, Sämtliche Werke nach Epochen seines Schaffens* (Munich: Hanser, 1989), vol. 12:48.
10. Goethe writes that the transition of form into another occurs "gleichsam auf einer geistigen Leiter, zu jenem Gipfel der Natur, der Fortpflanzung durch zwei Geschlechter." *Zur Naturwissenschaft überhaupt*, 30.
11. "Betrachten wir eine Pflanze insofern sie ihre Lebenskraft äußert, so sehen wir dieses auf eine doppelte Art geschehen, zuerst durch das *Wachstum*, indem sie Stengel und Blätter hervorbringt, und sodann durch die *Fortpflanzung*, welche in dem Blüten- und Fruchtbau vollendet wird. Beschauen wir das Wachstum näher, so sehen wir, daß, indem die Pflanze sich von Knoten zu Knoten, von Blatt zu Blatt fortsetzt, indem sie sproßt, gleichfalls eine Fortpflanzung geschehe, die sich von der Fortpflanzung durch Blüte und Frucht, welche *auf einmal* geschieht, darin unterscheidet, daß sie *sukzessiv* ist, daß sie sich in einer Folge einzelner Entwickelungen zeigt." Goethe, *Zur Naturwissenschaft überhaupt*, 65–66.
12. In this context one can consider the final lines of the elegy as the ascension of a ladder, a move coded in terms of the transcendence from the empirical to the spiritual through love: "Die heilige Liebe / Strebt zu der höchsten Frucht gleicher Gesinnungen auf, Gleicher Ansicht der Dinge damit in harmonischem Anschaun / Sich verbinde das Paar, finde die höhere Welt," Goethe, *Zur Naturwissenschaft überhaupt*, 76.

13. See the chapter titled "Das Pflanzenreich unserer Erde in Beziehung auf die Menschengeschichte," in *Ideen zur Philosophie der Geschichte der Menschheit*, in Herders Sämmtliche Werke (Berlin: Weidmannsche Buchhandlung, 1887), vol. 13:51–60.
14. "die Blüte, wissen wir, ist bei den Pflanzen die Zeit der Liebe. Der Kelch ist das Bett, die Krone sein Vorhang, die andern Teile der Blume sind Werkzeuge der Fortpflanzung, die die Natur bei diesen unschuldigen Geschöpfen offen darlegt und mit aller Pracht geschmückt hat," Herder, "Das Pflanzenreich," 53.
15. "Ihr großer Zweck sollte erreicht werden . . . dieser Zweck ist *Fortpflanzung, Erhaltung der Geschlechter*," Herder, "Das Pflanzenreich," 53–54.
16. It is worth recalling that *Erhaltung* has a double meaning in German: it can refer to the act of sustaining something, but also to the act of keeping something from falling apart or falling down, as one would support a child who stumbles. In an analogous sense, *Fortpflanzung* as *Erhaltung* is pursued in the face of potential risk: Herder's text identifies moments of potential loss, death, and discontinuity, all of which nature compensates for through sheer abundance, to the point of excess. "Die Natur braucht Keime, sie braucht unendlich viel Keime, weil sie nach ihrem großen Gange tausend Zwecke auf einmal befördert. Sie mußte also auch auf Verlust rechnen, weil alles zusammengedrängt ist, und nichts eine Stelle findet, sich ganz auszuwickeln," Herder, "Das Pflanzenreich," 54.
17. For another connection between teleological thinking and the discourse of *Fortpflanzung*, see also Kant's *Critique of the Power of Judgment* and the *Groundwork of the Metaphysics of Morals*. The "principle of teleology" in the *Critique of the Power of Judgment* (1790) clearly states "that in an organized being nothing that is preserved (*erhalten*) in its propagation! (*Fortpflanzung*) should be judged to be nonpurposive." Immanuel Kant, *Critique of the Power of Judgment*, trans. Paul Guyer and Eric Matthews (Cambridge: Cambridge University Press, 2000), 289. In this passage, Guyer and Matthews translate *Fortpflanzung* as "procreation." Whereas the emphasis is on *Fortpflanzung* as the procreation of the individual organism, Kant also invokes *Fortpflanzung* as both procreation and the propagation of a species. Already in the *Groundwork of the Metaphysics of Morals* (1785), for example, he formulates this tendency in broad terms with the claim that "Nature's end (*Zweck*) in the cohabitation of the sexes is *Fortpflanzung*, that is, the preservation of the species (*Erhaltung der Art*). Hence one may not, at least, act contrary to that end," *Groundwork of the Metaphysics of Morals*, trans. and ed. Mary Gregor (Cambridge: Cambridge University Press, 1996), 179. In each case there is an appeal

to necessity, whether understood as the execution of natural law or as an injunction not to act against it. There is also a parallel structure of ends and purposes. For Kant as well, what is preserved on the small scale, organism to organism, is the material counterpart to the concept of the species.

18. "Das Mittel, dessen sich die Natur hier [i.e., "zur Fortpflanzung der Gattung," J.H.] ebenso, wie allenthalben, zur Erreichung ihres Zweckes in freien Wesen bedient, ist ein natürlicher Trieb." Johann Gottlieb Fichte, *Das System der Sittenlehre nach den Principien der Wissenschaftslehre*, in *Fichtes Werke*, ed. Immanuel H. Fichte (Berlin: Walter de Gruyter, 1971), vol. 4:328.

19. Johann Gottlieb Fichte, *Foundations of Natural Right*, trans. Michael Baur, ed. Frederick Neuhouser (Cambridge: Cambridge University Press, 2000), 265. "Sollte sie möglich seyn, so musste die Gattung noch eine andere organische Existenz haben, ausser der als Gattung; doch aber auch als Gattung da seyn, um sich fortpflanzen zu können. Dies war nur dadurch möglich, dass die die Gattung bildende Kraft vertheilt, gleichsam in zwei absolut zusammengehörende, und nur in ihrer Vereinigung ein sich fortpflanzendes Ganzes ausmachende Hälften zerrissen würde. In dieser Theilung bildet jene Kraft nur das Individuum. Die Individuen, vereinigt, und inwiefern sie vereingt werden können, sind erst, und bilden erst die Gattung." Johann Gottlieb Fichte, *Grundlage des Naturrechts nach Principien der Wissenschaftslehre*, in *Fichtes Werke*, ed. Immanuel H. Fichte (Berlin: Walter de Gruyter, 1971), vol. 3:305–306.

20. "Versuch einer Geschichte der Schicksale der chemischen Theorie in den letzten Jahrhunderten." For a bilingual edition of each of these texts, see Jocelyn Holland, *Key Texts of Johann Wilhelm Ritter (1776–1810) on the Science and Art of Nature* (Leiden: Brill, 2010).

21. "Alles, was ist, erhält sich organisch. Jeder Stein entsteht in jedem Augenblick neu, erzeugt sich ins Unendliche fort. Nur sterben die Eltern des Kindes sogleich immer wieder, und so sieht man das Individuum nicht zunehmen. Kann man die Vernichtung aufheben, so schließt sich das neue an das fortbestehende Alte an, und nun hat wirkliche Fortpflanzung, Mehrung, statt." Ritter, *Key Texts*, 152–153.

22. In his reading of the *Oldest Document*, Ritter encounters Herder's description of a "really authentic old hieroglyph"—a kind of mystical pattern—contained in the first seven days of creation in *Genesis*. According to Herder, the hieroglyph in *Genesis* is not an arbitrary metaphorical image, but can provide an actual foundation for all branches of human thought. It is nothing less than a "primal image [. . .] according to

which gradually the entire script and symbolism of humans thus crafted so many inventions, arts and sciences" ["wie ihm etwa davon das *erste Urbild* worden, von dem man sich weiter *versucht*, an und nach welchem sich allmählich die *ganze Schrift* und *Symbolik* der Menschen also so viel *Erfindungen, Künste und Wissenschaften* gebildet?'"]. See Johann Gottfried Herder, "Älteste Urkunde des Menschengeschlechts," in *Werke*, ed. Günter Arnold, Martin Bollacher et al. (Deutscher Klassiker Verlag: Frankfurt am Main, 1989), vol. 5:269. Herder bases his discussion of the hieroglyph on an interpretation of the Egyptian God Theut as founder of the arts and sciences. Herder "reads" Theut in the hieroglyph through the first letter of his name (which Herder connects to the Greek letter "theta" or a symbol comprised of a circle with a line or cross in the middle) and makes a case for understanding Theut as monument, writing, and hieroglyph all at once. In the appendix to the fragment collection, Ritter translates this idea into the sphere of acoustics: "+ and O are directed towards the extremes of the *octave*,—which, in *language*, already expresses itself as the relation of the vowel to the consonant. And since only *both* together yield a *tone*, so does its cipher [i.e., a circle with the "X" inside, J.H.] contain even *plastic* significance." Ritter, *Key Texts*, 475.

23. "Jeder Ton ist ein *Leben* des tönenden Körpers und in ihm, was so lange anhält, als der Ton, mit ihm aber erlischt," Ritter, *Key Texts*, 476.
24. "Ganze Völkergeschichten, ja die gesammte Menschengeschichte, muß sich musikalisch aufführen lassen." Ritter, *Key Texts*, 476–478.
25. "Zeichnen Töne als bloße Bewegung, Bewegtheit, die bewegt, zuerst in Chladnis 'Klangfiguren' *sich selbst* analog auf und ein, so heißt Hören der Effekt des Sich-Eintragens der oszillatorischen Bewegung, die der Schall ist. Dem entspricht der Vorschlag Johann W. Ritters: 'Jeder Ton hat somit seinen Buchstaben immediate bei sich,' so daß wir 'überhaupt nur Schrift *hören*' und, daß wir 'lesen, wenn wir hören'—wie Ritter fortsetzt. Dies verschiebt aber das Konzept der Schrift und des Lesens, und zwar jenseits der Zeichenvermitteltheit, die diese *ausmacht*, zu einem Sich-Ein-Schreiben, als die Fortpflanzung und Wiedergabe von Bewegung wäre, und zur ungewußten 'Wahrnehmung' einer Eintragung, die kein Subjekt hat, eine Operation über Unterscheidungen, die [welche Unterscheidungen] kein Subjekt mehr, sondern die [Unterscheidungen] die Körper machen." Bettina Menke, "Töne—Hören," in *Poetologien des Wissens um 1800*, ed. Jochen Vogl (Munich: Fink, 1998), 75.
26. "Beym *Tone* übrigens, wo Discontinuität in der Zeit *doch* Continuität erzeugt, wäre am ersten zu fragen, ob in *bloßer Zeit aneinander grenzende*

differente Qualitäten. . . . nicht doch auch Processe, und Processe *eigner Art*, unter einander er zeugen mögen?—Schon in jeder *Fortpflanzung* von irgend etwas liegt dergleichen, da nemlich hier doch wirklich eine Zeit vergeht, bis der Process u. von *a* in *b*, von *b* in *c*, u. s. w. angekommen ist." Ritter, *Key Texts*, 496.

27. Ibid., 565.
28. Ibid., 623n.
29. Ibid.
30. Ibid.
31. Von Baer himself confesses to some uncertainty as to whether the speech was held in 1833 or 1834. In any case, it was published as part of an anthology titled, *Vorträge aus dem Gebiete der Naturwissenschaften und der Oekonomie* [Lectures from the Area of the Natural Sciences and Economics] (Königsberg: Unger, 1834).
32. "Die organischen Körper . . . zerstören sich selbst. Sie sind nicht nur steter Veränderung unterworfen, sondern ihre ganze Entwickelung ist ein Reifen zum Tode . . . mit dem Moment der Zeugung ist auch das Todesurtheil unterschrieben." Karl Ernst von Baer, "Das allgemeine Gesetz der Natur in aller Entwicklung," in *Vorträge*, 39.
33. "Der Naturforscher erkennt also, daß uns eine Nöthigung angeboren ist, an etwas Ewiges zu glauben, auch wo wir es nicht sogleich gewahr werden." von Baer, "Das allgemeine Gesetz," 40.
34. Von Baer states that "die Monade wie der körperliche Mensch nur wandelbare Erscheinungen sind, die aber während ihres flüchtigen, in Umgestaltung begriffenen Daseins die Keime zur Erneuerung desselben in anderen Individuen vorbereiten, die vorübergehenden Verwirklichungen eines bleibenden Gedankens, denn durch die Zeugungen hindurch wiederholt sich doch dieselbe Organisation . . . Die *Organisationsformen*, diese durch Zeugung zusammenhängenden Reihen, scheinen bleibende Gedanken der Schöpfung; die Individuen sind vorübergehende Darstellungen dieser Gedanken." "Das allgemeine Gesetz," 41.
35. "nichts ist im Manne vom Jünglinge, nichts im Jünglinge vom Kinde als eine Aehnlichkeit der Form und die Erinnerungen an frühere Erfahrungen, welche das Selbstbewußtsein zur Einheit des Ichs zusammenfaßt," von Baer, "Das allgemeine Gesetz," 41.
36. "Das Thier wird also durch Instinct zur Fortpflanzung genöthigt," von Baer, "Das allgemeine Gesetz," 47.
37. "so mußte er auch über die Zeugung gebieten können," von Baer, "Das allgemeine Gesetz," 47.

5

TREVIRANUS' *BIOLOGY*

Generation, Degeneration, and the Boundaries of Life

JOAN STEIGERWALD

The publication of a work entitled *Biology* by Gottfried Reinhold Treviranus in the first years of the nineteenth century has often been remarked in histories of the life sciences as one of several texts introducing the term *biology* in the years around 1800.[1] Treviranus' *Biology* was a large undertaking, appearing in six volumes—the first three volumes published from 1802 to 1805 and the remaining three, after a hiatus, from 1814 to 1822. Yet its significance has remained unclear, and it provides an ambiguous prototype for a science of biology. Certainly Treviranus did not establish a new discipline. He founded no school, enlisted no group of researchers, and created no institutional basis for a new approach to the study of living organisms. He did not instill a body of practices or techniques for the investigation of living organisms in his contemporaries, introducing no specific instruments, skills, or material dispositions enabling or delimiting biology as a distinct discipline. He did not instate a conceptual or analytic space encoded with characteristic functions, organizations, or normative judgments. He did not make a case for the utility of a science of life to social or political economies. His *Biology* did not become a widely cited textbook used for training physicians and naturalists, or instate a new genre of writing.[2] Indeed, Treviranus acquired no particular authority among his contemporaries. After receiving his doctorate from the University of Göttingen, the preeminent German medical school, he returned to his hometown of Bremen to become a consulting physician and professor at the local lyceum. His scholarly production was relatively modest, offering contributions to contemporary matters of concern—investigations of nervous forces, experiments on infusorians, and microscopic studies of the reproductive parts of invertebrates. His *Biology* did receive attention at the time, and was afforded lengthy, if mixed, reviews by widely read review journals such as the *Göttingische gelehrte Anzeigen* (Göttingen scholarly reports) and *Medicinisch-chirurgische Zeitung* (Medical-surgical periodical).[3] But in the years around 1800 it was difficult to define a science of biology, to determine what was meant by life, organic phenomena, or living organisms, and to specify methods for their investigation. The rapidly changing understandings and technologies of scientific inquiry, and changing relationships between different fields of inquiry, made it difficult to demarcate a distinct domain for the study of life. Works such as Treviranus' naming biology as a science of life did not establish a set of concepts or practices, but they did point to new problematics and concerns in the investigation of organic phenomena.

It is the first three volumes of Treviranus' *Biology* and its significance for understanding changes to studies of living organisms at the turn of the nineteenth century in the German context that is the focus of this chapter. In their content, these first three volumes display an evident tension between past studies in natural history and physiology, and a new synthetic science of life. In the lengthy introduction in the first volume, Treviranus laid claim to a scientific study of living organisms, modeled after philosophical conceptions of science as a system of knowledge, after theoretical rather than empirical approaches to medicine, and founded on certain principles. But in positing these principles for a science of biology Treviranus looked backward, not forward—backward to mid-eighteenth century speculations by Joseph Needham on organic matter and a vegetative force; and to earlier debates and speculations over *Lebenskräfte*. Yet Treviranus also pointed to new emphases in the study of living organisms—to a classification of the gradation of living organisms based on a comparative physiology; to studies of the influences of diverse environments on living forms; and to a history of living nature grounded in the history of the physical earth. It is this mixture of approaches to the study of living nature that made Treviranus' *Biology* an ambiguous exemplar for a new science of life.

What is most striking about Treviranus' work is his interest in the border zones of life. Treviranus was concerned with the boundary between the living and the lifeless, with what distinguishes vital from physical and chemical processes. He was also concerned with the boundaries between forms of life, particularly with the border zones between animals and plants populated with zoophytes, animal-plants (*Thierpflanzen*) and plant-animals (*Pflanzenthiere*), and contended that transformations in the forms of life are best studied in these border zones. The *Geschichte des physischen Lebens* (History of physical life), as Treviranus subtitled his work, is a story of the first origins of life and its progress into higher forms; it is also a story of degeneration and transformation, and of destruction and new generation. In the introduction to his *Biology*, Treviranus stressed the principles marking the boundary of living and lifeless nature—a viable matter and a *Lebenskraft* that counter the influences of the external world. But as he pursued the empirical part of his *Biology*, examining studies of the gradation, distribution, and history of living nature in extensive detail in the first three volumes, his interest in the boundaries of life led him not only to highlight the regularity and organization distinctive of living beings, but also to descend into the contingent, the deviant, and the material aspects of life. His studies of the generation and degeneration, especially in the lowest forms of life, increasingly blurred the boundary of living nature. Yet it was precisely in this focus on the border zones of life that Treviranus' work signaled the central problems for a new science of biology.

A SCIENCE OF LIVING NATURE: *BIOLOGY* AND THE BOUNDARIES OF LIFE

In setting out the characteristics marking biology as a science, Treviranus drew on contemporary debates in philosophy, medicine, and natural history. The full title of his book, *Biologie; oder Philosophie der lebenden Natur für Naturforscher und Aerzte* (Biology; or philosophy of living nature for naturalists and physicians), revealed the central significance of philosophy in this new science of life. The definition of biology, with which he opened his work, explicitly borrowed the language of contemporary German philosophy and of Immanuel Kant in particular. Biology was to take as its objects *"the different forms and appearances of life, the conditions and laws under which these states of affairs occur, and the causes through which they are effected."*[4] It was to be a science in the philosophical sense of a *Wissenschaft*—a theoretical and systematic approach to the study of life based on fundamental principles and laws.

Treviranus' specific model for science was Kant, who, however, he boldly corrected. Although only citing Kant's 1786 *Metaphysische Anfangsgründe der Naturwissenschaft* (Metaphysical foundations of natural science), he had clearly also read Kant's 1790 *Kritik der Urteilskraft* (Critique of the power of judgment), and found both inadequate. He took his lead from Kant's *Metaphysical Foundations*, which had constructed the principles of pure natural science corresponding to Newtonian mechanics by analyzing the concept of matter from contemporary empirical science through the pure concepts of the understanding.[5] But whereas Kant's analysis arrived at the fundamental principles of matter as the dynamic interplay between attractive and repulsive forces, Treviranus argued that only a single repulsive force is needed, if the repulsive force of each individual part of matter is understood as restricted through the repulsive forces of all the rest. Indeed, he represented the system of repulsive forces of the physical universe as an organized whole, in which each part is at once cause and effect, at once means and ends, using the terms through which Kant characterized living beings rather to characterize the physical universe.[6] Kant had contended in his third *Critique* that we cannot form a determinate conception of living organisms, and thus a science of life, but can only form a regulative idea of their apparent natural purposiveness through a limited analogy with our own purposive action. Treviranus argued instead that the conditions of life necessitated a second fundamental force, and argued further that he had a determinate conception of such a vital power or *Lebenskraft*. He thus brashly trampled Kant's careful epistemic arguments and critique of metaphysics.

Treviranus' *Biology* was one of several German texts appropriating the terms of philosophy and science to legitimate their approach to the study of life in the last years of the eighteenth century. Treviranus' own first contribution to physiology as a science concerned with the investigation of the causes and laws explaining the phenomena of life was an article "On the Nerve Power and its Way of Acting" in the second issue of the *Archiv für die Physiologie* (Archive for the physiology) founded by Johann Christian Reil in 1795.[7] Publishing in Reil's journal was an important opportunity for the young Treviranus, and the positive response to this first publication encouraged him to pursue the issues surrounding the possibility of a science of living nature. Reil used his prominence as professor of medicine and chief physician in Halle to attempt to establish a new foundation for physiology in Germany through his journal, by framing its aims and methods in the terms of Kant's critical philosophy in an extended article "On the Vital Power" in its first issue. He rejected Kant's restriction of the appellation *science* to mechanics, extending the term to encompass empirical inquiries such as chemistry and physiology. Yet, after Kant, Reil started from phenomena or appearances, arguing that natural science grounds particular appearances in particular qualities or forms and compositions (*Formen und Mischungen*) of matter. Powers (*Kräfte*), such as the *Lebenskraft*, he argued, are subjective concepts, the forms through which we think the relationship of appearances to the qualities of matter.[8]

Further proposals for a science of life appeared in the next years. In 1797 the Braunschweig physician Theodor August Roose prefaced his work on *Grundzüge von der Lehre von der Lebenskraft* (Fundamental traits of the doctrine of the vital power) by representing the work not only as a contribution to science (*Wissenschaft*) but also as an outline of a biology.[9] The medical instructor and writer Karl Friedrich Burdach also used the term *biology* in his guide to academic lectures *Propädeutik zum Studium der gesamten Heilkunst* (Propaedeutic to the study of the whole of the healing art) in 1800. His medical textbook provided an encyclopedia of the healing arts divided into distinct sciences, from natural history, chemistry, and physics to anatomy, physiology, psychology, pathology, and medical treatments. The term *biology*, however, was reserved for physiology as the appearances or doctrine of life of the human being regarding "its form, or its composition or its characteristic powers."[10] If thus widening the concept of medicine as a science to include areas of empirical study such as anatomy and natural history, Burdach retained a traditional notion of physiology as the theoretical part of medicine concerned with the forces or causes of the phenomena of life.[11]

The appearance of several texts attempting to define a science of life or biology at the end of the eighteenth century might seem to suggest the beginning of the formation of a discipline. These developments seem to complement the formation of chemistry and physics as disciplines distinguished by their own communities of scholars with their own journals and textbooks, and the introduction of the natural sciences as areas of instruction in universities distinct from the faculties of philosophy or medicine in the German context.[12] But such formations remained precarious and ambiguous, especially in the areas concerned with the study of living organisms. Professors often held chairs, both simultaneously and sequentially, in disparate areas, and often taught courses listed in different faculties from their appointments. Physiology was taught in faculties of medicine and philosophy; natural history was taught in faculties of medicine, cameral science, or under the category of natural science; and chemistry was a topic of instruction in several faculties. The fact that articles on contentious matters such as organic vitality and vital powers were published in journals of chemistry and physics as well as medicine and physiology highlights the lack of a distinct domain for the study of life at the time. The flourishing of scientific and natural historical societies in the years around 1800 also provided important communities for individuals with an interest in the study of nature, but such civic societies often fostered a wide range of pursuits. Indeed, the *Gesellschaft Museum* in Bremen was formed in 1783 through merging a literary society and a society for physics and natural history. It was a burgeoning society in 1797 when Treviranus arrived, with more than 200 members, but the society was dominated by merchants, lawyers, and bankers, not scholars.[13] Treviranus did not find a model for a science of life in university curriculum, journals, or scientific societies.

Nevertheless, the framing of studies of living organisms in terms of a science and the language of philosophy was widespread. Such a framing can even be found in Christoph Wilhelm Hufeland, the prominent Weimar physician and professor of medicine at the University of Jena, who argued for the importance of basing medical practice in experience rather than on theoretical systems. Hufeland founded the *Journal der Practischen Heilkunde* (Journal of practical medicine) in 1795, the year Reil founded his *Archive*, to promote the practical healing arts as a counter to the excessive number of theoretical medical journals in German. Yet in the same year he also published *Ideen über Pathogenie und Einfluss der Lebenskraft auf die Entstehung und Form der Krankheiten* (Ideas on pathogeny and influence of the vital power on the origin and form of illnesses), which presented a medical and pathological theory premised on the principle of a special vital power as the cause of both health and disease, and which he enlisted to police the boundary between living organic beings and dead nature.[14]

In the introduction to his *Biology* Treviranus, like Hufeland, argued for a vital power as the principle founding biology as a science and as the condition of making life possible. He thus rejected the Kantian critical strictures that Reil had attempted to establish in physiology, and accepted the *Lebenskraft* as an objective natural power rather than as a subjective concept aiding judgments of the relationships of appearances to the qualities of matter. Treviranus' science of biology was premised on a division of nature into the living and the lifeless, and on the importance of boundaries [*Gränzen*] between them, and the need to demarcate a distinct domain of inquiry into the question "What is life?" Yet he also recognized the difficulty of separating vital phenomena from physical and chemical phenomena, given that all matter in the universe is organized and active.[15] His definition of life acknowledged this tension: The distinctive character of life is the similarity or uniformity (*Gleichförmigkeit*) of appearances with dissimilar or contingent influences (*ungleichförmige oder zufällige Einwirkungen*) of the external world.[16] The continual stimulus of external influences is as necessary to the activities of life as is their capacity to resist them. But the *Lebenskraft* polices the boundaries between them, to prevent the living organism from succumbing to the vortex of natural activity. Any weakness in the *Lebenskraft* results in a breach of this boundary and deviations from the healthy course of life, if not its destruction.[17]

In arguing for a *Lebenskraft*, Treviranus was enlisting arguments for a principle of life that had preoccupied German physicians and naturalists a decade earlier. Principles of life and *Lebenskräfte* proliferated in German language publications in the early 1790s. Several authors of tracts in medicine, physiology, and natural history drew a boundary around organic nature in the context of perceived intrusions from new studies in chemistry and physics. The event stimulating this intensity of attention to *Lebenskräfte* was the publication of Christoph Girtanner's "Abhandlungen über die Irritabilität, als Lebensprincip der organisirten Natur" ("Treatises on Irritability, as the Principle of Life in Organized Nature") in 1791.[18] Girtanner argued not only that irritability was the basis of organic vitality, but also that it could be understood chemically drawing on Antoine Lavoisier's new chemical theory making oxygen the active element in combustion and respiration. Endorsing French chemistry at a time when prominent German chemists were still repudiating it was already controversial, but to suggest a chemical basis of life was too controversial for most German physicians. Soon numerous reviews, articles, and books appeared presenting competing arguments regarding the principle of life, some defending sensibility and others irritability but most advocating for a *Lebenskraft* that demarcated the living from the nonliving. Naturalists also participated in the debate, most notably Kielmeyer, then professor of zoology and chemistry and curator of the natural history museum at the *Karlsschule* in Stuttgart, who combined interests in physiology with natural history in comparative studies of organic vitality. In a widely cited lecture from 1793, Kielmeyer contended that each living organism is constituted through a unique interrelation of vital powers, adding generative powers to those of irritability and sensibility.[19] The flurry of publications concerned with a principle of life peaked in 1795, when it received one of its most sophisticated treatments by Hufeland. Treviranus' first physiological publications were contributions to these debates, and in *Biology* he continued to position his arguments for a *Lebenskraft* in relationship to notions developed in the context of these controversies from the early 1790s—notions of excitability, receptivity to stimulus, and the capacity of living organisms to resist the effects of the physical world.[20]

But in the second half of the 1790s the bald positing of special vital powers to police the boundary between living and lifeless nature was increasingly criticized as German physicians and physiologists used new tools from chemistry and physics for the investigation of organic processes. Reil launched his *Archive* at least in part to counter the recent proliferation of *Lebenskräfte*. In his article "On the Vital Power" he argued that each living organism, and indeed each organ, had a unique organic form and composition (*Form und Mischung*) transmitted through the generative matter in reproduction; it is the specific organic forms and compositions that give rise to the specific processes of life, from irritability to generation, and these should be studied through anatomical and chemical investigations. If we are able to know the form and composition of matter underlying organ appearances more precisely, Reil contended, the concept of a special vital power is no longer required.[21] Alexander von Humboldt proffered a similar position. Although he initially appealed to a *Lebenskraft* to account for the unique capacities of living beings, by the late 1790s his extensive galvanic experiments and his investigations of the excitability of organic parts and the effects of chemicals and gases on organic processes led him to align his position with that of Reil, and to conclude that vital functions are complex alterations in the form and composition (*Form- und Mischungsveränderungen*) of organic matter.[22] Reil and Humboldt encouraged the investigation of these complex organic processes in their relationships to chemical and physical processes, troubling the boundary between the organic and inorganic even as they remained convinced of its presence. By 1802 Treviranus had clearly moved away from Reil's position. Despite his interest in the border zones between the living and the nonliving, despite his conception of life as requiring both the stimulus of nonliving processes and their resistance, he criticized the tendency of "the present age" to reject the *Lebenskraft* and to replace it with "the mere form and composition of matter."[23]

Instead, Treviranus turned to a previous age, to the mid-eighteenth century, to elaborate his conception of life. After Joseph Needham and Georges-Louis Leclerc, Comte de Buffon, he argued for the existence of an organic matter through which all living beings possess life.[24] In support of this claim, Treviranus enlisted Needham's infusion experiments from 1750, in which Needham purported to have demonstrated the formation of simple plants or animals from decayed organic matter. Finding infusions teeming with microscopic moving bodies, even in those that had been boiled and sealed, Needham concluded that they were the product of organic matter and its vegetative force rather than airborne seeds. Treviranus' former professor at the University of Göttingen, Henri-August Wrisberg, had confirmed Needham's experiments, and had published these results in 1765. In the second volume of *Biology*, Treviranus examined the infusion experiments of Needham and Wrisberg in detail, countered the experiments of Lazzaro Spallanzani and other critics of Needham's results, and presented his own experimental findings in support of Needham.[25] Such experiments, he concluded, confirmed the existence of an organic or a viable matter—a *"lebensfähige Materie"* or "matter capable of life."

In his introduction to *Biology* Treviranus posited the principles of his system of biology—the interplay between viable matter, the *Lebenskraft* and external influences in the formation of living nature—that would inform the detailed studies of the subsequent volumes. Indestructible and indecomposable, unalterable in its essence and present since the first formation of the universe, he claimed that viable matter is formless or lacking organization yet takes constant forms through the persistent influence of external causes. He further claimed that viable matter is inseparably bound with the *Lebenskraft*, which ensured the similarities of appearances of living organisms in the face of contingent and changing external influences. The *Lebenskraft* draws a partition between the living body and the rest of nature, making the organization and purposive relations of the parts of the living organism more excellent than that of lifeless bodies. Viable matter takes specific determinate forms through its connection with external stimuli whose influences are filtered through the vital power specific to each organic form, which resists certain external influences and allows others. But under changing physical conditions and thus new external influences viable matter can take on new forms. The higher the forms of life, the farther away is the boundary of the arbitrary effects of external nature on that living form, yet the simple forms of life are capable of being continually formed from viable matter under the appropriate external conditions. Treviranus stressed the importance of the reciprocal interaction of viable matter and the vital power, for if the *Lebenskraft* alone was the cause of living organisms no new forms of life would be possible and death would be a transition to lifeless nature. But death does not involve a transition to lifeless matter, but instead through viable matter is a transformation into other forms of life.[26]

Treviranus' introduction offered a summary review of different possible systems of biology—life as a capacity of matter, as dependent on a vital power, or as the reciprocal interaction between viable matter and a vital power—before deciding on the last system. His review has striking similarities to a review of three possible principles of life by Friedrich Wilhelm Joseph Schelling in his 1799 *Erster Entwurf eines Systems der Naturphilosophie* (First outline of a system of the philosophy of nature)—a material principle, a vital power, or a principle derived from both. Schelling also invoked a concept of excitability similar to that of Treviranus to account for the capacity of living organisms to preserve their own sphere of activity and prevent their assimilation into the larger natural world; each organism remains receptive to the stimulus of external material influences but also engages in activities resisting them.[27] Yet Treviranus summarily dismissed Schelling for confusing the living with the ensouled, not engaging Schelling's natural philosophy in a sustained examination, and using it only to position his *Lebenskraft* between spirit and matter and in a domain specific to the living.[28] Schelling, however, appealed to the antique figure of a world soul not to postulate a hyperphysical spirit, but quite the contrary to indicate an embodied spirit or vitalized matter. In fact Schelling, writing as a natural philosopher immersed in post-Kantian critical philosophy, rejected the appeal to an immaterial *Lebenskraft* as an objective power to police the boundary between the living and lifeless as uncritical. He rather regarded the difference between the inorganic and organic as a difference of degree of activity and organization rather than a difference in kind, and used "boundary concepts" such as excitability to interrogate the differences and relationships between them.[29] Despite Treviranus' rejection of Schelling's philosophy of nature and the differences in their approaches, both conceived life through its boundary with the inorganic, regarded a descent into organic matter the key to understanding the generative capacities of living organisms, and allowed for transformations of living forms through the intrusion of the material and contingent.

BIOLOGY AS THE HISTORY OF PHYSICAL LIFE: GENERATION AND DEGENERATION

In turning from the philosophy of life to the empirical details of the history of physical life, Treviranus made manifest the scope of his conception of biology as a science of living nature that encompassed natural history, physiology, geography, geology, and related fields of inquiry. Treviranus opened his discussion in volume one by reminding readers of his concern with the "borders of living nature," and the relationships of living nature with the lifeless. He then proceeded to voluminous presentations of the classification and gradation of living beings based on a comparative physiology; of the distribution and variations of living beings based on the physical and geographical characteristics of the world; and of the alterations of living beings in relation to historical changes to the earth. The last two volumes, published in 1814 and 1818, would present the functions of nutrition and chemical processes, movement and irritability, and intelligence and sensibility. The first three volumes, published from 1802 to 1805, however, foregrounded the generative and degenerative capacities of living nature as the basis for understanding the different forms and variations of living beings. In the accounts of the gradation, distribution, and history of living nature in these volumes, the simple organisms whose generation Treviranus claimed to be demonstrated in infusion experiments were accorded a central place.

In the first section of his history of physical life, Treviranus rehearsed the critique of traditional natural history as a merely descriptive and artificial register that had become commonplace in the latter eighteenth century, arguing instead for a history of nature.[30] Kant had made an important argument for a history of nature in his 1775 treatise *Von den verschiedenen Racen der Menschen* (Of the different human races):

> The history of nature (*Naturgeschichte*), of which we are still almost entirely lacking, would teach us about the changes in the earth's form, including those the earth's creatures (plants and animals) have undergone through natural migrations, and their degenerations thereby from the original form of the stem genus.

Kant's history of nature was to be founded on "Buffon's rule, that animals that produce fertile young with one another belong to one and the same physical species [*Gattung*]."[31] Buffon's multivolume *Histoire naturelle* would become one of the most influential publications of the eighteenth century; in opening its first volumes in 1749 with a general history of nature and a theory of generation, Buffon was proposing a new conceptual framework for understanding the diversity of species grounded in genealogical relationships. If initially rejecting a transformation from one species into another, arguing that species descend from the original progenitors of a first creation and that a relative constancy of form is secured by interior molds, Buffon increasingly broadened his notion of propagation to include wider degrees of material relationships and variations effected through changed climate, soil, and diet.[32] Building on Buffon's work, in his treatise on race Kant offered a theory of generation and degeneration based on inherited germs and predispositions lying dormant in the generative material until particular environmental conditions induced an unfolding; but once unfolded dispositions became permanent racial characteristics, and passed on through reproduction. By his 1790 *Critique of the Power of Judgment* Kant found a new model in Johann Friedrich Blumenbach. Blumenbach's essay *Über den Bildungstrieb* (On the formative drive) appeared in multiple editions throughout the 1780s and 1790s, to become the most widely cited work on generation in the latter eighteenth century in the German context, commended for presenting a convincing demonstration of epigenetic formation. Like Buffon, Blumenbach made his account of generation central to his works on natural history, works in which he speculated on radical revolutions in the earth's past leading to the extinction of extant life. But Blumenbach restricted the creation of new species of organic beings to a divine hand, and attributed the propagation of species to the formative drive of the generative matter unique to each living kind. The limitation of the degenerative effects of physical factors on species was key to his arguments regarding the different races, yet unity, of humankind.[33] Despite significant differences in the details of their natural histories, Buffon, Kant, and Blumenbach all contributed to the development of a history of nature, with both the capacity for degeneration and its constraint effected through generative processes. Their works were widely discussed by German naturalists at the end of the eighteenth century, and Treviranus was familiar with their contributions. But they did not provide the basis for his history of physical life; Treviranus offered a far more radical account of the transformations produced by the generation and degeneration of living nature.

Natural history in the eighteenth century extended beyond concerns with defining species or races and the question of their historical transformation. It was tied to the project of the management of the natural economy and the progress of the state, to promises of improvement and prosperity, and accordingly had broad social and political implications. It was also tied to the building of collections, of state museums and botanical gardens, and to the development of networks of exchange of information and specimens, and accordingly to imperial enterprises. Such projects encouraged an interest in the effects of changed climates and the potentials of cultivation on varieties of plants, animals, and human beings. Collections and exchanges of specimens in effect provided laboratories for natural history, with collections allowing the comparisons of species from different parts of the world and exchanges of living specimens allowing experiments in transplantation and acclimatization. Naturalists such as Buffon and Blumenbach, located in important research centers in Paris and at the University of Göttingen, were able to base their work in the history of nature in growing collections and networks of exchange.[34] As a student in Göttingen, Treviranus had the opportunity to examine Blumenbach's natural history collection; the Bremen *Gesellschaft Museum* also had an extensive natural history collection. Although Treviranus would later produce respected studies on the anatomy of the reproductive organs of fish, amphibians, and small rodents as well as invertebrates, his early research focused on simple organisms.[35] Such research appears in his *Biology*. But he based his history of physical life largely on the expertise of others, drawing on an extensive literature on classification, comparative anatomy, comparative physiology, geography, and geology.

In the first volume of *Biology* Treviranus provided a classification of living organisms in a graduated series, beginning with man and mammals, and descending to the lower forms of life. He rejected classification systems based on single parts, but he also did not enlist Buffon's genealogical rule, which indeed many naturalists had critiqued as impractical for the concrete work of identifying and ordering species. Rather he argued that the biologist must look "at the similarity or difference of the entire organization," and combine considerations of the composition, structure and function of parts.[36] Blumenbach had made the total relation of characters of the "entire *habitus* of the animal" the key to his taxonomic system, basing his natural history upon comparative physiology as well as comparative anatomy. In his *Handbuch der Naturgeschichte* (Handbook of natural history), Blumenbach opened his presentation of each class of animal with a discussion of their primary physiological functions—from reproduction and nutrition to respiration, movement and nerve functions—as well as discussion of their external characteristics and abode.[37] Treviranus also foregrounded comparative physiology in his classification system. He compared the brain, nervous system, and sensory organs, the heart and lungs, the digestive system and sexual organs—tracing these systems not only through a graduated series of animals, but also noting analogous systems in zoophytes and plants. But the principles of his system of classification were not clearly articulated, and in its details his natural history depended on the descriptions of established experts in specific classes of organisms. In the end, he distinguished particular species mainly through descriptions of particular parts, instead of presenting a new system founded in an integrated biological organization as promised. By 1805, in volume three of *Biology*, however, Treviranus singled out generation or propagation as the most important characteristic of living nature, and proposed the different modes of generation or propagation as the basis for a new classification of the different kinds of living organisms.[38]

Important for the whole of Treviranus' *Biology* was his demarcation of a distinct domain of zoophytes next to the kingdoms of plants and animals. These consisted of simple organisms that had no consistent placement in contemporary classification systems, but which Treviranus demarcated as a border zone between animals and plants. They included what he termed animal-plants (*Thierpflanzen*) or *Zoophyta*—such as infusion animals, corals, hydra, sea anemones, and sea stars—and plant-animals (*Pflanzenthieren*) or *Phytozoa*—such as algae, fungi, lichen, and seaweeds.[39] These simple organisms are the central topic of volume two of *Biology*, which details the distribution of living nature; here Treviranus argued that zoophytes are found and indeed are continually being generated throughout the world, and provided the demonstrations of their generation in infusion experiments. They are also central to volume three, which sketches the revolutions of earth and the first forms of life to appear in the history of the earth; Treviranus argued that zoophytes are the original forms or *Urkeime* from which all other forms of living nature are generated through transformations and degenerations.

The second volume of *Biology*, appearing in 1803, was concerned with the present condition of the earth and the distribution of living nature across all areas of the globe. Treviranus was interested in how not only the physical features of the earth but also geographical zones affected the distribution of plants, zoophytes, and animals—from land masses, seas and mountains; through rivers, deserts and swamps; to the latitude and dimensions of a region. A substantive part of this volume was simply a register of the different species found in different regions of the earth, drawn from the writings of Linnaeus, the Forsters, Peter Simon Pallas, and others who had traveled to these regions. Emphasizing that all parts of the earth are home to living creatures, he recounted alleged observations of worms, insects, and even amphibians in sealed caves, the interior of trees, or blocks of marble. But zoophytes were given special significance, because of their presence everywhere—they are found on the tops of the highest peaks, in the depths of the sea and in the darkest caves; in the inhospitable polar lands and in the heat of the desert; and in sulphur lakes and in boiling springs.[40] This ubiquitous presence of zoophytes was used to support his claim of the capacity of new forms of life continually to be generated from the viable matter. Indeed, in the text Treviranus moved from detailing the widespread distribution of the different forms of life to experimental demonstrations of the formation of infusorians from dead and decaying organisms. But according to the principles outlined in his introduction, viable matter only takes determinate forms through the influence of external causes. He thus moved on to examine new physical and chemical studies of the effects of atmospheric gases, of climate, of temperature, water and light, and of atmospheric pressure and electricity on living organisms, to account for the degeneration of the different kinds of living organisms in these different regions. His conception of life as under the continual stimulus of influences of the external world provided the basis for a history of life that included the kind of physical geography that would become central to the work of Humboldt and others in the course of the nineteenth century. Living organisms are everywhere, he argued, and the history of physical life is continual cycles of generation and degeneration, death and new generation, through the interplay of viable matter and physical influences.

In 1805, in the third volume of *Biology*, Treviranus turned his attention to the past. Here he examined contemporary theories of geological change and of rock formations or layers, following Humboldt in appealing to both sedimentation and the internal heat of the earth to account for such changes. Given the emerging picture of an earth in endless transformation, he concluded it is "necessary to show that through these transformations also living nature must be changing."[41] He appealed to new studies on the temporal sequence of geological layers and the fossilized forms of life associated with these, outlining a history of the revolutions of the earth and of life on earth that Blumenbach and Georges Cuvier had begun, and that would become central to biology in the nineteenth century. He also examined debates over the interpretation of fossils and the extinction of species, and determined that the evidence for extinction, at least in the case of higher organisms, was decisive. For Treviranus the central event in this history of the earth, however, was the first emergence of life, and in this event zoophytes and infusorians again took central place. Following the historical stages of formations of the rocks introduced by Abraham Gottlob Werner—primitive, transitional, layered (*Flötz*), and more recent alluvial and volcanic, classes of rocks or layers—he placed the first appearance of life, of simple *Thierpflanzen* and *Pflanzenthieren*, in the transitional stage of formation.

In contrast to his introduction to *Biology* in 1802, Treviranus now proposed an origin of life from the materials of the earth, but did so by making life an attribute of the whole world. It was the *Lebenskraft* as a fundamental power alongside the repulsive force, he argued, that made life possible. Through the *Lebenskraft*, in the earliest periods of the earth's history, the carbons, metals, and earths necessary for life were first synthesized, and then later these were synthesized into simple living forms. Once living beings appeared, they began to further transform the world, changing its airs, waters, and earths, producing new materials that would then combine into new living forms. Whereas in 1802 Treviranus had foregrounded the border between living and lifeless nature, he now argued that this opposition is present only from our viewpoint, but not for nature: "All, the universe itself, possesses life." "The first origin of life in general thus loses itself in the origin of the universe."[42] The separation of the living and the lifeless only emerged as individual living organisms separated from the earth and formed into small self-enclosed worlds. The consequence of the universal presence of the *Lebenskraft* as a fundamental power of nature is that the origin of life is not a singular and contingent event. The first appearance and transformations of living forms do not only follow revolutions of the earth, but also occur independently and indeed continuously. Treviranus allowed that fossils of sea life found in the middle of continents and on the tops of mountains could be attributed to a massive flooding of the earth, and that some migrations of organisms to new regions do occur. But he contended that most fossils are of living forms indigenous to the area in which they are found. He concluded that the origin of life was widespread across the regions of the earth. Moreover, arguing for analogies between changes in the present and the past, he contended the emergence of life is ongoing. Treviranus' arguments in the third volume of *Biology* were tied to those in the second volume for the presence and formation of simple living organisms, of zoophytes, everywhere and at all times.

The gradation of living forms detailed in volume one was thus given a historical basis in volume three, with the higher forms of life emerging from the lower forms, from *Thierpflanzen* and *Pflanzenthieren*, in Treviranus' history of physical life. Treviranus even included human beings in this history of life. He contended that humans and apes only emerge after the alluvial period, although he also admitted gaps in our knowledge as to the exact period in which humans were formed and living nature approached its present form. But he also presented evidence of the transformation or degeneration of human beings over time, and posited that such transformation would continue and that nature had not reached its highest level yet in humans. "All on earth is fluid and fleeting, the kind like the individual, the species like the kind. Even the human being perhaps will some day pass and be transformed."[43]

To account for these transformations of living nature, both those past and those ongoing, Treviranus enlisted processes of degeneration and generation. The effects of the physical and chemical influences of the external world in the degeneration of living nature recounted in volume two were also given a historical dimension in volume three. Together with the geological revolutions and the changes wrought through the emergence of living nature—changes affecting the physical earth, its climate and materials—Treviranus argued that over time such external influences transformed the organization of living beings, and resulted in the disappearance of some kinds and the appearance of new ones. He posited that "each kind, like each individual, has certain periods of growth, flourishing and death, but that its death is not dissolution, as with the individual, but degeneration."[44] Only changes to the whole of nature could produce substantive changes to species, with changes to local conditions affecting only individuals. But, again, zoophytes form an exception; susceptible to any alteration of external conditions, they are continually undergoing transformations. Although Treviranus contended that degenerations are the principle cause of the deviations of the diverse forms of living nature from the forms of their ancestors, he also suggested that processes of generation play a role in producing transformations. He offered an extended discussion of different modes of propagation in volume three, offering through these a new classification of the different kinds of living being. Although he acknowledged a widespread role for sexual reproduction, even in animal hermaphrodites and plants, he challenged claims for its ubiquity, pointing to the many simple animals and plants in which no evidence of sexual organs or sexual reproduction can be found. Once again it is the zoophytes that are significant here, and Treviranus added his own study of the reproduction of *confervae*, a kind of algae, to other studies of their distinctive modes of reproduction.[45] He was particularly interested in the capacities of certain kinds of organisms, especially *Thierpflanzen* and *Pflanzenthieren*, to propagate in different ways at different times or under different circumstances, seeing in such variations a possibility for transformation. After examining Joseph Gottlieb Kölreuter's experiments with hybrids from the 1760s, which demonstrated a gradual reversion to the type of the mother or father after a few generations, he concluded interbreeding is not a potential source of the transformation of species. But he did allow that misbirths and monstrosities, and especially the inheritance of deviant forms, were productive of transformations of the kinds of living beings. In analogy to the degeneration of living nature under alterations of the conditions of the earth, he argued misbirths are caused through changed conditions, through accidental physical influences or ill conditions of the generative matter. Through his account of generation and degeneration Treviranus claimed "to have solved one of the most difficult problems of biology."[46]

CONCLUSION

Treviranus offered an account of the continual degeneration of living nature under the influences of the external world, one far more radical than those proposed by Buffon, Blumenbach, or Kant. What he did not offer was an explanation of how, in such a vision of continual change, higher organisms capable of complex functions of life arise from simple zoophytes, or indeed any organized being is generated from unorganized viable matter. Nor did he explain how, despite continually changed conditions, the distinct characteristics of species and races, or indeed individuals, are inherited or propagated across generations. These, in contrast, were central concerns of Buffon, Kant, and Blumenbach. In invoking a *Lebenskraft* to provide this explanation, but without offering a detailed account of how it acted to form living organisms beyond bald claims of its capacity to filter external influences and synthesize the materials necessary for life, he was identifying rather than resolving a central problem of biology.

Moreover, by volume three of *Biology* Treviranus had reconceived the boundary between living and lifeless nature, presenting a bold vision not only of the continual transformation and degeneration of living nature, but also of its first formation and constant new generation. In his 1802 introduction, the *Lebenskraft* polices the boundary of life, ensuring the regularity of the appearances of living nature by resisting the contingent influences of the external world. Viable matter is also present from the first formation of the universe, taking determinate forms through the interaction of the *Lebenskraft* and specific physical influences. But in volume three, completed only three years later, the *Lebenskraft* was reconceived as working with lifeless nature to form the materials necessary for life. Not only do living beings dissolve into viable matter to be regenerated into new living forms, but simple forms of life are continually emerging from physical materials. Treviranus' *Biology* was highly speculative, even for the time, despite his amassing of supportive citations. Kant, with his usual critical caution, had censured such speculative histories, arguing for a restriction of the possible forms of generation and transformation to those we can presently observe. Treviranus observed no such strictures, boldly engaging in a speculative archaeology regarding ancient and radical transformations in the forms of life. He offered an audacious narrative sketching how living organisms arose from the "womb of [the] common mother" to ascend to higher and diverse forms.[47] But it is also a story of a descent into the material basis of life, and the degenerations and transformations of living forms occurring under contingent external influences. Despite his introductory criticisms of the present age for its focus on alterations in the form and composition of organic matter, in its subsequent volumes Treviranus' *Biology* gathered together influential new studies of the material conditions of life. Indeed, his preoccupation with the border zones of life, with the boundary between living and lifeless nature, and with the simplest forms of living organisms, led him to explore the interactions between the materials of life and its physical environment, and the generation and degeneration of diverse forms of life in suggestive new ways.

Biology brought together new directions in the study of life, in its emphasis on comparative physiology, the distribution and history of living nature, and the interaction of the forms life with their physical environment. But its contribution to these new studies was muddied by Treviranus' insistence upon concepts that had already met with substantive critiques, on a *Lebenskraft* and viable matter. Certainly his work did not establish a new discipline or a science of life; its significance lies in recognizing and naming a developing new approach to the study of living nature—biology. But if pointing to a new domain for scientific inquiry, Treviranus' *Biology* complicated the demarcation of the boundaries of this domain by focusing upon the interactions and continuities of living and lifeless nature. The significance of the text might best be regarded as marking the tensions and ambiguities of first attempts to articulate a science of biology in the years around 1800.

NOTES

1. See, for example, Brigitte Hoppe, "Le concept de biologie chez G.R. Treviranus," in *Colloque international "Lamarck,"* ed. Joseph Schiller (Paris: Librarie Scientifique et Technique A. Blanchard, 1971), 199–237; Timothy Lenoir, *The Strategy of Life: Teleology and Mechanics in Nineteenth-Century German Biology* (Chicago: University of Chicago Press, 1982); Joseph A. Caron, "'Biology' in the Life Sciences, a Historiographical Contribution," *History of Science* 26 (1988):223–268; Peter McLaughlin, "Naming Biology," *Journal of the History of Biology* 35 (2002):1–4; and Kai Torsten Kanz, " . . . die Biologie als die Krone oder der höchste Strebepunct aller Wissenschaft," *"NTM: Zeitschrift für Geschichte der Wissen Technik und Medizin"* 14 (2006):77–92. The most extensive work on Treviranus remains the excellent dissertation by Timothy F. de Jager, *G.R. Treviranus (1776–1837) and the Biology of a World in Transition* (Unpublished doctoral thesis, University of Toronto, 1991). The paper here is indebted to Jager's work, if offering a different reading of Treviranus' significance.
2. See Michel Foucault, *Discipline and Punish: The Birth of the Prison,* trans. Alan Sheridan (New York: Vintage Books, 1995); and Timothy Lenoir, *Instituting Science: The Cultural Production of Scientific Disciplines* (Stanford: Stanford University Press, 1997).
3. *Göttingische gelehrte Anzeigen* no. 96 (June 16, 1804), 951–960; no. 205 (December 26, 1805), 2041–2048; and *Medicinisch-chirurgische Zeitung* 1 (1803), 33; 1 (1804), 81; and 4 (1805), 225.
4. Gottfried Reinhold Treviranus, *Biologie: oder, Philosophie der lebenden*

Natur für Naturforscher und Aerzte (Göttingen: Johann Friedrich Röwer, 1802–1822), vol. 1:4; emphasis in the original.
5. Kant, famously, argued in the *Metaphysical Foundations* that in any doctrine of nature there can be only as much proper science as there is mathematics therein, but Treviranus did not address this particular demand. Immanuel Kant, *Metaphysische Anfangsgründe der Naturwissenschaft*, in *Kants gesammelte Schriften* (Berlin: Königlich Preussische Akademie der Wissenschaften, 1902–1983), vol. 4:470.
6. Treviranus, *Biologie*, vol. 1:25–43.
7. Gottfried Reinhold Treviranus, "Ueber Nervenkraft und ihre Wirkungsartm," *Archiv für die Physiologie*, vol. 1, 2 (1796):8.
8. Johann Christian Reil, "Über die Lebenskraft," *Archiv für die Physiologie* 1, 1 (1795):8–162. Reil was introduced to Kant's critical philosophy through his contact with the Kantian Marcus Herz in Berlin before his move to Halle. See Thomas H. Broman, *The Transformation of German Academic Medicine, 1750–1820* (Cambridge: Cambridge University Press, 1996), 86–88; and Robert J. Richards, *The Romantic Conception of Life: Science in the Age of Goethe* (Chicago: University of Chicago Press, 2002), 252–261.
9. Thomas A. Roose, *Grundzüge der Lehre von der Lebenskraft* (Braunschweig: Thomas, 1797), iii.
10. Karl Friedrich Burdach, *Propädeutik zu Studium der gesammten Heilkunst* (Leipzig: Breitkopf Härtel, 1800), 62.
11. Andrew Cunningham, "The Pen and the Sword: Recovering the Disciplinary Identity of Physiology and Anatomy Before 1800. I: Old Physiology—The Pen," *Studies in History and Philosophy of Biological and Biomedical Sciences* 33 (2002):631–665.
12. Karl Hufbauer, *The Formation of the German Chemical Community* (Berkeley: University of California Press, 1982); William Clark, "German Physics Textbooks in the Goethezeit," *The British Journal for the History of Science* 35 (1997):219–239 and 295–363; Thomas Bach and Olaf Breidbach, "Die Lehre im Bereich der 'Naturwissenschaften' an der Universität Jena zwischen 1788 und 1807," *NTM: Zeitschrift für Geschichte der Wissen Technik und Medizin* 9 (2001):152–176; and Paul Ziche, "Von der Naturgeschichte zur Naturwissenschaft: Die Naturwissenschaften als eigenes Fachgebiet an der Universität Jena," *Berichte zur Wissenschaftsgeschichte* 21 (1998):251–263.
13. Andreas Schulz, "'. . . Tage des Wohllebens, wie sie noch nie gewesen' Das Bremen Bürgertum in der Umbruchszeit 1789–1818," *Historische Zeitschrift* 14 (1991):25–45. See Denise Phillips, *Acolytes of Nature:*

Defining Natural Science in Germany, 1770–1850 (Chicago: University of Chicago Press, 2012).
14. *Ideen über Pathogenie und Einfluss der Lebenskraft auf Entstehung und Form der Krankenheiten als Einleitung zu pathologischen Vorlesungen* (Jena: Academische Buchhandlung, 1795); and Christoph Wilhelm Hufeland, *Journal der practischen Arzneykunde und Wundarzneykunst* 1, 1 (1795):iii. See Broman, *The Transformation*, 104–118; and Klaus Pfeifer, *Medizin der Goethezeit: Christoph Wilhelm Hufeland und die Heilkunst des 18: Jahrhunderts* (Weimar: Böhlau, 2000).
15. Treviranus, *Biologie*, vol. 1:3–6 and 16–18.
16. Ibid., 23 and 38.
17. Ibid., 51–52 and 59–79.
18. Girtanner's treatise appeared first in the widely read French journal *Observations sur la physique* in 1790 and then in German translation in the *Journal der Physik* in 1791. Christoph Girtanner, "Abhandlungen über die Irritabilität als Lebensprincip in der organisierten Natur," *Journal der Physik* 3 (1791):317–351 and 507–537.
19. Carl Friedrich Kielmeyer, *Über die Verhältnisse der organischen Kräfte*, ed. Kai Torsten Kanz (Marburg an der Lahn: Basilisken-Presse, 1993).
20. Gottfried Reinhold, *Physiologische Fragmente*, 2 vols. (Hannover: Hahnische Buchhandlung, 1797–1799); and Treviranus, *Biologie*, vol. 1:59–70.
21. Reil, "Über die Lebenskraft," §§1–7.
22. Compare Humboldt *Aphorismen aus der chemischen Physiologie der Pflanzen*, trans. Gotthelf Fischer von Waldheim (Leipzig: Voss, 1794) and *Versuche über die gereizte Muskel- und Nervenfaser, nebst Vermuthungen über den chemischen Prozess des Lebens in der Thier- und Pflanzenwelt* (Berlin: Rottman, 1797), vol. 1:126; vol. 2:41–52 and 430–436.
23. Treviranus, *Biologie*, vol. 1:52.
24. On Buffon's conception of organic molecules, see Florence Vienne's chapter in this volume. Treviranus focused on Needham's work. Oken, whom Vienne also discusses, draws on Treviranus as well as Buffon. See Joan Steigerwald, "Degeneration: Inversions of Teleology," in *Marking Time: Romanticism and Evolution*, ed. Joel Faflak, Naqaa Abbas and Josh Lambier (Toronto: University of Toronto Press, forthcoming).
25. Treviranus, *Biologie*, vol. 2:267–295 and 319–352. On Needham's experiments, see John Tuberville Needham, "A Summary of Some Late Observations upon the Generation, Composition, and Decomposition of Animal and Vegetable Substances," *Philosophical Transactions* 45 (1748):615–666; and Shirley A. Roe, "John Tuberville Needham and the

Generation of Living Organisms," *Isis* 74 (1983):159–184. Spallanzani's work appeared in German in 1769: Lorenzo Spallanzani, *Physikalische und Mathematische Abhandlungen* (Leipzig: Gleditschens, 1769).

26. Treviranus, *Biologie*, vol. 1:59–70, 97–103; and vol. 2:264–267, 353 and 403–404.
27. Compare Treviranus, *Biologie*, vol. 1:83–103; and Friedrich Wilhelm Joseph Schelling, *Erster Entwurf eines Systems der Naturphilosophie*, in *Historisch-kritische Ausgabe*, ed. Jörg Jantzen et al. (Stuttgart: Frommann-Holzboog, 1976–), vol. I, 7:117–131 and 170–184.
28. Treviranus, *Biologie*, vol. 1:33. In contrast, Treviranus' brother Ludolf Christian, who studied in Jena and attended Schelling's lectures, was quite sympathetic to Schelling's natural philosophy. See Brigitte Hoppe, "Ludolf Christian Treviranus," in *Naturphilosophie nach Schelling*, ed. Thomas Bach and Olaf Breidbach (Stuttgart: Frommann-Holzboog, 2005), 737–773.
29. Schelling, *Von der Weltseele*, in *Historisch-kritische Ausgabe*, ed. Jörg Jantzen et al. (Stuttgart: Frommann-Holzboog, 1976), vol. I, 6:82 and 253–257, and Schelling, *Erster Entwurf*, 127–128.
30. Chapters in this volume by Robert Bernasconi, Florence Vienne, Jocelyn Holland, and Staffan Müller-Wille offer excellent analyses of shifting attitudes to the history of nature, generation, and degeneration, propagation of kinds, and race and heredity in the latter eighteenth century that provide the backdrop to Treviranus' work.
31. Immanuel Kant, *Von den verschiedenen Racen der Menschen*, in *Kants gesammelte Schriften* (Berlin: Königlich Preussische Akademie der Wissenschaften, 1902–1983), vol. 2:434 and 429. See Robert Bernasconi's chapter in this volume.
32. See in particular Buffon's 1753 essay on the domestic donkey, in Georges-Louis Leclerc, Comte de Buffon, *Natural History*, trans. William Smellie, 20 vols. (London: T. Cadell and W. Davies, 1812), vol. 4:165.
33. Johann Friedrich Blumenbach, *Über den Bildungstrieb und das Zeugungsgeschäfte* (Stuttgart: Gustav Fischer Verlag, 1971); *Beyträge zur Naturgeschichte* (Göttingen: Dieterich, 1790), 24–25; *Handbuch der Naturgeschichte*, 4th ed. (Göttingen: Dieterich, 1791), §1; and *The Anthropological Treatises of Johann Friedrich Blumenbach*, trans. Thomas Bendysche (London: Longman, Green, Longman, Roberts & Green, 1865).
34. Emma C. Spary, *Utopia's Garden: French Natural History from Old Regime to Revolution* (Chicago: University of Chicago Press, 2000); Anke te

Heesen and Emma Spary, eds., *Sammeln als Wissen: Das Sammeln und seine wissenschaftsgeschichtliche Bedeutung* (Göttingen: Wallstein, 2001); John Gascoigne, "Blumenbach, Banks, and the Beginnings of Anthropology at Göttingen," in *Göttingen and the Development of the Natural Sciences*, ed. Nicolaas Rupke (Göttingen: Wallstein Verlag, 2002), 86–98; Staffan Müller-Wille, "Figures of Inheritance, 1650–1850," in *Heredity Produced: At the Crossroads of Biology, Politics and Culture, 1500–1870*, ed. Staffan Müller-Wille, Hans-Jörg Rheinberger (Cambridge, MA: MIT Press, 2007), 177–204; and Mary Terrall, "Following Insects Around: Tools and Techniques of Eighteenth-Century Natural History," *British Journal for the History of Science* 43 (2010):573–588.
35. P. Smit, "Treviranus, Gottfried Reinhold," in *Dictionary of Scientific Biography*, ed. Charles Coulston Gillespie (New York: C. Scribner, 1976), vol. 8:260–262.
36. Treviranus, *Biologie*, vol. 1:160–162.
37. Johann Friedrich Blumenbach, *Handbuch der Naturgeschichte*, 1st ed. (Göttingen: Dieterich, 1779), 56–57; and *Anthropological Treatises*, 188–190. See Peter McLaughlin, "Blumenbach und der Bildungstrieb: Zum Verhältnis von epigenetischer Embryologie und typologischem Artbegriff," *Medizinhistorisches Journal* 17 (1982):362–363.
38. Treviranus, *Biologie*, vol. 1:155–174; and vol. 3:229–365.
39. Treviranus classified the *Zoophyta* as *Asteriae, Actininiae, Pennatulae, Corallia, Gorgoniae* and *Infusoria*; and the *Phytozoa* as *Fungi, Confervae, Fuci, Lichenes, Hepaticae, Musci, Filices* and *Naiades*. Treviranus, *Biologie*, vol. 1:405–425.
40. Treviranus, *Biologie*, vol. 2:3–30.
41. Ibid., vol. 3:8.
42. Ibid., 39–40.
43. Ibid., 226, 22–23, and 166–173.
44. Ibid., 225–226.
45. Ibid., 314–323.
46. Ibid., 462.
47. Ibid. Compare Kant's arguments in the *Kritik der Urteilskraft*, in *Kants gesammelte Schriften* (Berlin: Königlich Preussische Akademie der Wissenschaften, 1902–1983), vol. 5:427–428; and his reviews of Herder's *Ideas*, "Immanuel Kant Recensionen zu J.G. Herders Ideen zur Philosophie der Geschichte der Menschheit," in *Kants gesammelte Schriften* (Berlin: Königlich Preussische Akademie der Wissenschaften, 1902–1983), vol. 8:44–66.

Part II

ARTICULATIONS OF RACE
AND GENDER

6

SKIN COLOR AND THE ORIGIN OF PHYSICAL ANTHROPOLOGY (1640–1850)

RENATO G. MAZZOLINI

Since the second half of the nineteenth century, the historiography of physical anthropology has concentrated on identifying the chronological sequence of the variously influential studies that gave rise to the systems used to classify the human races. Derived from this endeavor is an emphasis placed by the bulk of this historiography on the emergence of the concept of race. Concomitantly, this notion of race was arbitrarily extended in works by scholars belonging to historical periods in which the term (or any equivalent notion) did not even exist. On the other hand, even the most cautious historiography—that which has sought to understand the political and social implications of racial classifications—has been to some extent enthralled with the concept of race, both in its wholesale reduction of physical anthropology to racial classification alone and in its historical assessment of it as nothing more than colonial ideology, thereby relegating important issues to the status of a pseudoscience and, indeed, favoring their deceitful concealment.

The insistence with which both the apparently neutral historiography, and that critical of the concept of race have looked at early racial classifications, is not without its intrinsic interest. It is my belief, however, that both historiographies do not yield understanding of the various scientific, religious, political, and cultural reasons for the complex genesis of physical anthropology as a discipline, as well as the meaning it has assumed in the various contexts of European society and culture in the course of the last five centuries. On the basis of wide-ranging research conducted over many years, it is my intention in this chapter to suggest some elements for a general rethinking of the origin of physical anthropology during its gestation between approximately the mid-sixteenth century and the mid-nineteenth century. However, rethinking this historical phase requires us, first of all, to consider it not as a "phase" that inevitably produced the racial theories of the nineteenth century but rather to examine it in relative autonomy and detached from its presumed future. In other words, when we read the texts of the seventeenth and eighteenth centuries, which a well-established historiographical tradition regards as constituting the foundation of physical anthropology, we must set this interpretation aside and instead ask what problems these texts addressed and sought to explain. We must also ask what categories Europeans employed to classify non-Europeans before the advent in the nineteenth century of such notions as *race, racism, civilization, primitive, monogenesis, polygenesis*, and so on. We must enquire, for example, as to whether physical otherness was effectively a cognitive problem in the sixteenth and seventeenth centuries. Questions of this kind undermine our relative certainties on the perceptive modes of Europeans in that period, and they require a systematic reinterpretation of the documentation available to us.

If we take physical anthropology to be that area of scientific research that concerns itself with human biological diversity, then we must concur with those who locate the flourishing of the discipline in the second half of the eighteenth century, when scholars like Buffon (1749) called it the "natural history of man."[1] If instead we ask when, in Western Europe, human biological diversity became a problem to explain and why it became a problem, we find ourselves in great difficulties, because the question has not been subjected to systematic historical inquiry. Moreover, the belief in the existence of men with the heads of dogs and tails, which from Herodotus through Pliny and Pomponius Mela was still widespread in the Middle Ages and the early modern age, can hardly be taken to be the matter of an alleged physical anthropology, because the difference was not ascertained. It was not the object of research, nor was it thematized. This is not to imply that the belief is not of historical significance; only that it cannot be taken to be part of an explicit anthropology or "natural history of man." The historical importance of the belief resides in the fact that scholars of antiquity and of the Middle Ages were ready to recognize beings with the heads of dogs and tails as human beings. Consequently, although these scholars located these beings at the margins of the explored world, their physical diversity did not overawe them, and they were willing to accept a more polymorphous image of man as human. Since the sixteenth century this image has grown increasingly restricted in its range.

It was perhaps because the travelers of the late fifteenth and sixteenth century did not come across men with dog heads and tails during their long, dangerous, and exhausting journeys that their accounts of these travels say little about the physical features of the peoples that they met along the way and, instead, abound with descriptions of customs and beliefs. These accounts had a forceful impact on the imaginations of those who read them in Europe, depicting as they did forms of behavior and beliefs at odds with those permitted in closely structured and disciplined European society. They also prompted the first collection of observations that became the stock in trade of a new discipline: ethnology. But at that time, they were rather the object of theological and moral reflection, which gave rise to speculative models of another sort of humankind: "man in the state of nature" in the seventeenth century and the "noble savage" in the eighteenth. The activation of an imaginative capacity now able to conceive a different kind of human existence was not a matter of minor importance. It is comparable in its implications and effects on European culture to the great sixteenth- and seventeenth-century debates on the structure of the cosmos. This imagination, however, closely interwove with another dimension: the past. Historically erudite inquiries conducted by using the method of historical derivation sought to legitimate vested power on genealogical grounds, propounding the unlikely descendents of coeval noble houses from Roman or even Trojan families. This gave rise to renewed interest in Greco-Roman civilization, which was proposed as the model of enduring civil and artistic virtues against which the present—the sixteenth and seventeenth centuries—should measure itself in an ideal endeavor to emulate and surpass it, and of which the *querelle des anciennes et des modernes* was only one of the many symptoms. However, seeking to emulate a beloved and mythicized past also meant emulating its decadence. Europeans were thus foreseen a fate similar to the one that befell the Greco-Romans, which the millenerianist Christian tradition viewed as even more likely, and indeed inevitable. Anxiety about the future and a desire, almost always unexpressed, for endless conservation prompted for some a comparative concern with the relative features and merits of the various European populations. This generated, on the one hand, a peculiar genre of essays on "national characters," which enjoyed great success in the eighteenth and early nineteenth centuries and, on the other, a new view of history in which the Christian idea of salvation was secularized into that of progress.

When Western European scholars realized that the Americas formed a continent unknown to the ancients and separated from the Old World by interminable expanses of ocean, the fact that it was inhabited by humans became an intractable theoretical problem for them. How could human beings have reached the Americas? This was the question. It obviously did not concern a physical difference among humans; but concerned, rather, their different location in a space where the Old World was deemed the center. The reason why this constituted a problem was the narrative of the origin of the human species and its distribution as portrayed in the book of Genesis. This narrative was considered a revealed truth, and for Christians it constituted a binding framework for interpretation of humanity as a whole and therefore also for the inhabitants of the New World. Numerous theories of migration contended that a population from the Old World had settled—after traveling by sea or by land—in the West Indies. The new themes propounded in numerous works of the sixteenth and seventeenth centuries were those of latent analogies between the rituals of the Amerindians and those of the populations of the Old World. But the explanatory paradigm for the spatial distribution of mankind was migration and colonization by a population stemming from the Old World; that is, by predecessors of the same Europeans who were now once again migrating and colonizing. Only by hypothesizing ancient migrations, in fact, could the presence of a human species on an incredibly distant continent be explained. This was a self-projection through time, and no European could possibly conceive that the Earth might have been populated from the Americas. The notion of migration accounted for the diversity of the places inhabited by man and gave reassurance that the human species was one and one alone. And it is a postulate that remains today.

In its chronological distribution, from the first fifteenth-century accounts to the studies of the late seventeenth century, the literature on the Amerindians displays an evident shift of interest among Europeans from an intellectual curiosity about their customs and beliefs and also their origins, to an almost exclusive insistence on the latter. It is as if the Amerindians, with their diversified and mysterious cultures, disappeared from the scene—and their more complex political and social structures did indeed effectively disappear—so that what remained was their geographical distribution, which proved so difficult to fit into the framework of classical historiography and biblical revelation. Consequently, the theory propounded by Isaac La Peyrère (1655) of a twofold creation (one of all men, the so-called pre-Adamites, and of Adam, the progenitor of the Jewish people alone), whereby the Amerindians were pre-Adamites who had survived the Great Flood, attracted harsh criticism from both Protestants and Catholics.[2] It was attacked because it emphasized the inherent contradictions in the Mosaic revelation (who, in fact, were the people who tilled the land at the time of Adam, and who was Cain's wife?), and undermined the canonical beliefs that underpinned the Europeans' conception of the origin of the human species and its distribution around the globe. It reinforced the idea that there might exist a humanity parallel and alien to the one elected by God and the sole object of sacred history and gave rise to a dichotomous view of humanity, which was powerful and selective. There were the descendants of Noah (the second Adam) on the one hand, and, on the other, everyone else; namely those apparently excluded from the restricted paradigm of mankind's origins from Adam and Eve. Although assailed by the proponents of universal evangelization, this view took deep root, becoming the covert criterion—especially in everyday life—which justified exclusion, subjugation, and exploitation on historical grounds.[3]

The geographical expansion of the Europeans furnished further elements and suggestions for recasting traditional notions, including those concerning the difference between men and animals and, in particular, between men and monkeys. That the anatomical structure of men and apes displayed surprising similarities had been well-known since antiquity. Indeed, Galen's anatomy—as critically pointed out by Vesalius and the Italian anatomists of the sixteenth century—was much more an anatomy of the ape than of the human. However, while the anatomists of the early modern age boasted that they had emancipated human anatomy from that of apes, news arrived of creatures similar to monkeys whose appearance and behavior closely resembled human beings. The information on the intellectual abilities of certain apes—like the Asiatic "pithecanthrope," which "could be mistaken for a wild man" (Gesner 1551); the industrious ape (probably gorilla), which built itself shelters against the rain (Battel 1613); the emotions displayed by the female orangutan (de Bondt 1658); the anatomy of a chimpanzee, which had thirty-four features in common with other apes and forty-eight with the human species (Tyson 1699)—induced some naturalists to reconsider the similarities found between men and apes and, more generally, man's place in the animal kingdom.[4] While Descartes postulated an unbridgeable difference between humans and animals, whereby the former were endowed with reason and were therefore spiritual, and the latter were machines entirely devoid of reason, some naturalists were so struck by the parallels between physical features and behavioral patterns that they claimed that the intellectual differences between men and apes were not qualitative, but merely a matter of degree and that they could be attributed to a different conformation of the brain.⁵ From the point of view of zoological classification, for Linnaeus (1735), humans should be placed together with apes in the order of "Anthropomorphs"—that is, what were later termed "Primates."⁶ "I ask you," Linnaeus wrote in 1747 to one of his correspondents, "and the whole world [if there is] a difference of genus according to the principles of natural history between man and ape. I certainly did not know of any."⁷

The debate on the relation between humans and apes, man's place in the animal kingdom, and his alleged singularity, grew more intense. It mingled with other interconnected themes like the notions of "species" and "reproduction," or the reciprocal links among organic structure, function, and environment. The debate thus influenced the systems of classification employed and the images of nature and man that were the products and/or matrices of those classifications. The solution of one problem, therefore, entailed changes or adjustments to contiguous or more general theories. The discussion was not restricted to specialists like Buffon, Linnaeus, Camper, Blumenbach, Lamarck, and Cuvier alone, but involved a large part of European culture in an overall rethinking of the order of animate nature. Indeed, the issues, which preoccupied successive generations of naturalists from circa 1749 onward, had an interest of general scope: What was the natural order, and if one existed, how could it be perceived or conceived? Inevitably, questions of this level of importance were influenced by the aspirations, projects, experiences, and reflections relative to another order—the political and social system—which in those same years, amid reforms, revolutions, wars and restorations, envisaged or established new relationships among men. Due to often simultaneous correspondences, the two orders of phenomena (i.e., the natural and political-social) overlapped in the minds of even those who sought to keep them distinct, giving rise to a slow but inexorable shift from classificatory criteria, in which the spatial dimension and the ontology of beings predominated, to criteria in which the temporal or historicist dimension and the becoming of beings held sway.

"WHY ARE THE ETHIOPIANS BLACK?"
SKIN COLOR AS A RESEARCH OBJECT

The process just outlined becomes more evident if we consider the debates on black skin in their chronological sequence from the mid-seventeenth to the mid-nineteenth century. At first sight, it may seem curious that, as the Europeans continued their geographical expansion, the main physical difference

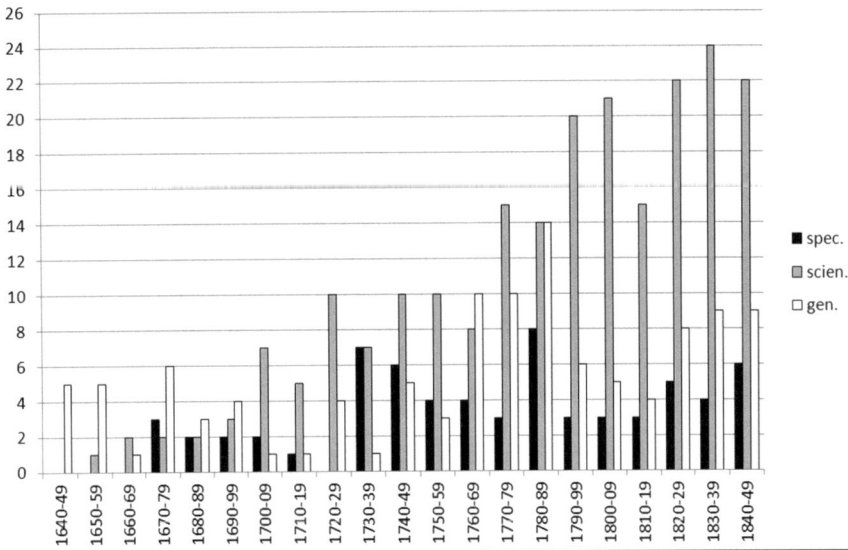

Chart 6.1

noted was not a difference perceived in populations previously unknown to them, but a difference perceived—and which therefore became problematic—in populations with which they had been acquainted since time immemorial. And yet, numerous seventeenth-century texts show unequivocally the presence of a recurrent question: "Why are the Ethiopians black?" At that time, and in the two following centuries, *Ethiopian* was the learned term used to denote the sub-Saharan African. In the seventeenth and eighteenth centuries, the question gave rise to a large number of monographs or specialist articles. The issue was also addressed in chapters or sections of manuals on anatomy, physiology, natural history, natural philosophy, general history, theology, and in travel literature. From the second half of the eighteenth century onward, it was a central topic of texts on the natural history of man.

Chart 6.1 shows the chronological distribution, between 1640 and 1849, of the first editions of four hundred works that either sought to explain why the skin color of sub-Saharan Africans was black or used skin color as a criterion for the classification of humankind. Although the chart refers to research still in progress, and is therefore incomplete also because it does not consider entries in dictionaries and encyclopedias, it nevertheless demonstrates that the problem of the pigmentation of Africans was a major and constant subject of inquiry. The chart shows the "peaks" of the debates in the specialist literature (spec.)—which only includes works whose title refers to black skin—as well as the presence of the topic in scientific handbooks (scien.) and more general works (gen.).

The chart acquires greater significance if we consider that, during the period in question, not a single work whose title referred to the skin color of the Chinese or Amerindians was published. There is one exception, but it is easily explained. The controversial book by Cornélius de Pauw, *Recherches philosophiques sur le Américains*, contains a chapter entitled "De la peau des Américains," but this in fact describes the skin of Africans, not Amerindians![8] The obsession of Europeans with the skin color of Africans, moreover, is evidenced by further findings. For the period 1675–1810 documentation exists on at least thirty-eight dissections of Africans, the sole intention of which was to study the color of their skin. This is a substantial number considering that, in the same period, there is no evidence of a single dissection of an Amerindian or an Oriental for the same purpose.

But what matters do all these books and articles discuss? Most of them conduct anatomical analysis on the location of pigmentation in a particular layer of the skin and offer physiological hypotheses and speculations on the theological and physical causes of such pigmentation. They sometimes draw up a classification of human peoples and their temperament; or they discuss the origin of humankind and the unity of the human species on the basis of skin color, speculating as to Adam's color. Some texts set out aesthetic judgments and discuss the intellectual capacities of Africans, while others deal with such enigmatic and disconcerting phenomena as albinism among individuals whose parents were black. For some authors it represented a return by blacks to humankind's original coloring. It is somewhat surprising to find that, although the Europeans had displayed great curiosity in the albinism of Africans since the sixteenth century, it was not until the late eighteenth century that they realized that albinos existed among themselves as well.[9] This suggests that they previously had been unaware of them.

While European scholars sought to explain why the Ethiopians were black, some authors raised questions concerning the coloring of animal pelts and the changes that take place in them from one season to the next. Others instead examined the way in which skin color was transmitted to the next generation by crossbreeding among whites, blacks, and Amerindians.[10] In this manner, new areas of research were opened up; for instance, on the coloring of animals (on which no contemporary history of significance exists) or on what some historians call "pre-Mendelian genetics." In this regard, maximum importance should be given to a paper by William Charles Wells, published in 1818, in which the author developed a theory of natural selection while conducting research to explain the skin color of Africans.[11]

As regards anatomical research, it should be pointed out that before the development of the cell theory in the 1830s and the use of the microtome to prepare animal tissue for microscopic examination, skin was analyzed anatomically not with vertical sections but by inspecting layers detached from each other by means of boiling or slow maceration. These techniques enabled Marcello Malpighi to separate the horny and the "mucous" (also termed *reticular membrane*) layers of the epidermis and to suggest, in 1665, that color was located in the mucous layer.[12] In 1677, by dissecting the skin of an African woman, Johannes Pechlin managed to show that the "black pigment" (i.e., granules of melanin) was contained in the Malpighian layer; not in the horny one. Pechlin's location of skin color was verified and accepted by the great majority of subsequent authors.[13] However, in the late seventeenth and early eighteenth century, Malpighi's conjecture and Pechlin's findings were misinterpreted (among other reasons) because, from a technical point of view, it was easier to discern the Malpighian layer in Africans than in Europeans. Consequently, a number of authors claimed that the former had an extra "cuticle" with respect to the latter. The erroneousness of this theory was proved mainly by two brilliant Dutch anatomists, Ruysch and Albinus, who demonstrated the existence of a Malpighian layer in Europeans as well.[14] The difference consisted solely in the color of the mucous layer—just as Malpighi had already suggested in 1665 without having performed any dissection on Africans, but by basing his conclusion on a comparative analysis of animal skins.

If the literature used to draw up chart 6.1 is assessed as a whole, from a positivist point of view, the main results obtained by anatomists and physiologists during the period considered can be summed up as follows. Their merit was to:

(a) consider blackness a natural rather than a metaphysical phenomenon;
(b) correctly locate it in the Malpighian layer of the skin;
(c) show that the dermis is white in both Europeans and Africans, and that the skin of both comprises a differently colored Malpighian layer;
(d) demonstrate that the "black pigment" (termed *æthiops animal* in 1765 by Le Cat, and *melaina* by Bizio in 1825) was not present solely in the skin, but also in part of the eye and in cuttlefish ink;[15]

(e) identify the melanin granules as the cause of hair color;
(f) develop the theory that the skin was a selective barrier between the external environment and the internal environment of the body;
(g) demonstrate, after much speculation, that albinism is a pigmentary disorder common to all humans;
(h) formulate what has to date been the greatest paradox relative to the skins of whites and blacks: that a dark skin absorbs heat, which may not be required in a warm climate, while a light skin reflects heat, which might be useful in a cool climate.

EXPLANATIONS OF BLACKNESS

The anatomical difference sought between Europeans and Africans thus turned out to be only a difference of coloration in a layer of the epidermis present in all human beings. For many scholars this feature was to be explained by singling out its causes. It should be noted that it was almost never asked why the Malpighian layer of Europeans was whitish. It was the norm from which those who were not white deviated. For that matter, Adam was generally imagined as white. As a consequence, the search for the causes of the pigmentation of the sub-Saharan Africans concerned only the latter.

Numerous causes were considered and discussed. They can be distinguished between metaphysical (or, theological) causes and physical causes. The former were looked for in holy scripture. A number of authors argued that Africans were the descendants of the cursed people of Canaan and that their skin was dark because of Noah's curse.[16] Others argued that Africans were the descendants of Cain; that blackness was the mark impressed upon Cain and that they were survivors of the Deluge.[17] Neither explanation was backed by any textual evidence, but scholars endeavored to support them with elaborate biblical exegesis and by citing Hebrew and Arabic texts.

The physical causes examined were: climate, sperm, bile, blood, the conditioning of the fetus by the mother's imagination, purported chemical substances produced by the body and not excreted through respiration, the nervous fluid, and an ancient pathology (leprosy). Each author proposed one or several causes. The significance of each theory—as well as its political and social implications—can only be assessed by setting it in the historical context of those who formulated it. However, at a general level, it may be of interest to observe the fortunes of two theories on the basis of the works surveyed to construct the histogram in chart 6.2.

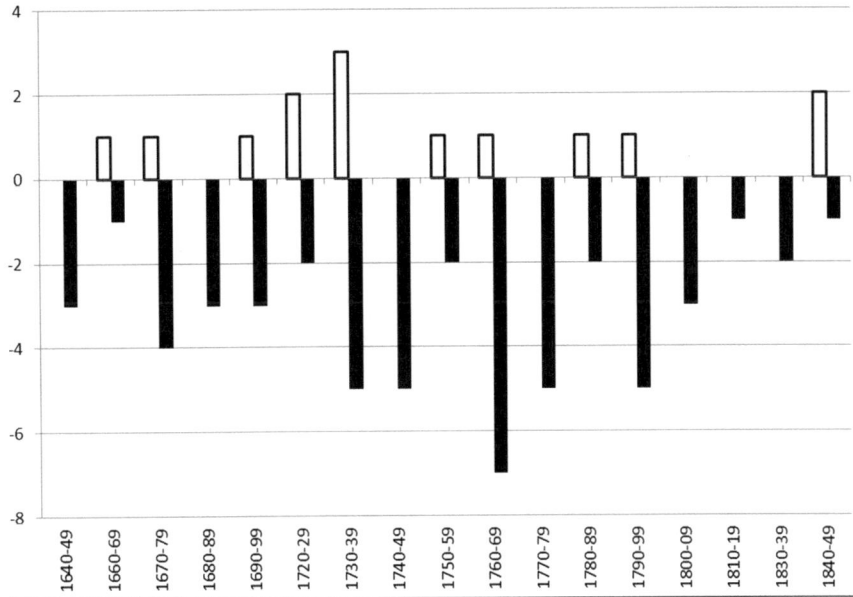

Chart 6.2

The above histogram shows the temporal distribution of works that offered theological explanations for the blackness of Africans and also of those that denied the validity of these explanations. It shows that the theological explanation was more contested (-) than it was asserted (+). Nevertheless, its large number of adversaries, as well as the number of references made to it, suggests that it was the opinion most widely held by the general public. Moreover, the reason why so many naturalists dismissed it was its reliance on a metaphysical (and not a natural) cause. Overall, the histogram shows the decline—only among scholars—of a supposed biblical paradigm that allegedly explained the difference of the sub-Saharan Africans.

The histogram in chart 6.3 shows the fortunes of the climatic theory. First developed in antiquity, this theory offered a naturalistic explanation of human pigmentation and was reformulated in the eighteenth century. It was based on two groups of observations: those relative to the tanning of the human skin, and those that pointed out that human peoples were distributed from the poles to the tropics according to a scale of colors, which ranged from light to dark. Since this explanation postulated that humans would change their skin color if they lived at different latitudes for a few generations, it was easily contradicted by many travelers of the sixteenth century, who found

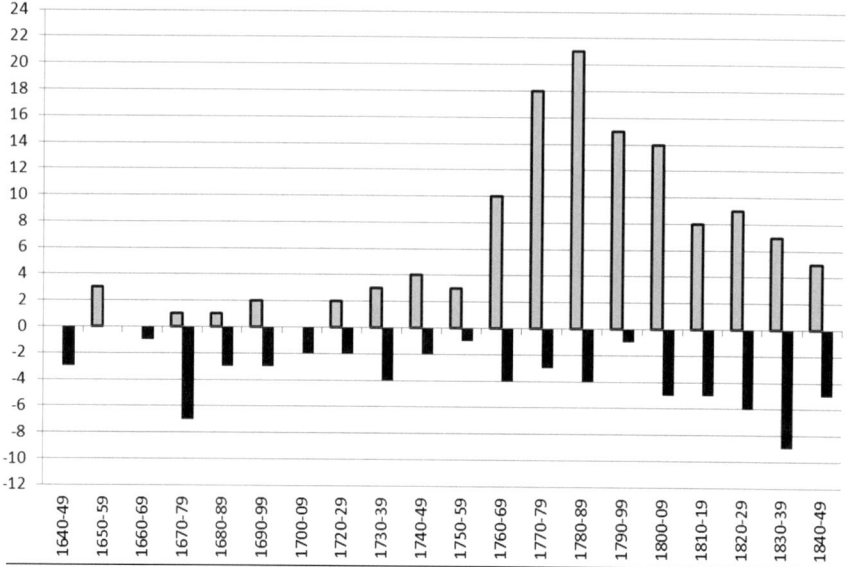

Chart 6.3

no change of skin color among the descendants of Africans who had been transported to different latitudes. Despite the criticisms brought against the theory, when appropriately recast, it enjoyed great success in the second half of the eighteenth century. This was principally because it found such authoritative proponents as Montesquieu and Buffon, but it was once again discredited in the mid-nineteenth century.

Buffon considered the natural history of man to be the history of the human species through time, and he argued that the effects of the climate-environment could not be ascertained in only a few generations. To give credibility to his theory, he requested that much longer time spans should be considered. This was similar to the request made by the late nineteenth-century evolutionists, who asked for more time for the slow mechanisms of evolution to become apparent.

SKIN COLOR, CLASSIFICATION, AND DIFFERENCE

In the course of their inquiries into the color of Africans, some authors drew dramatic comparisons of color, mucous layer, temperament, and mind. The core of the connection was what is known today as *cutaneous sensitivity*.

Then, as today, the majority of physiologists agreed that the horny layer of the skin is not sensitive and that tactile sensations are aroused by pressure on the papillary body of the skin (discovered by Malpighi in 1665). However, some authors argued that the pigment present in the mucous layer of Africans impeded delicate sensation and consequently hampered the harmonious activity of the mind. William Frederick Van Amringe, for example, in a bulky work entirely devoted to the natural history of man, *An Investigation of the Theories of the Natural History of Man*, published in New York in 1848, wrote that the structure of the skin,

> [M]ust necessarily modify the functional power of the nervous system, and therefore affect the quality of impressions upon the brain in a very high degree. The difference, in this respect, in the species of man, independently of capacity, or form of the skull, is amply sufficient to produce all the differences observable in all the species. It will account for the *strenuous* temperament of the Shemites [or white species]; the *passive* temperament of the Japhetites [or yellow species]; the *callous* temperament of the Ishmaelites [or red species]; and the most hopeless and lowest of all, the *sluggish* temperament of the Canaanites [or black species]; —whose only hope for an ameliorated condition appears to lie in the bondage incident to a "servant of servants."[18]

The above passage provides an example of the value judgments inscribed in most classifications of the human species based on skin color. Contrary to what one might suppose, these classifications were relatively new among Europeans. In the sixteenth and early seventeenth centuries, the large groups of populations distributed around the globe were not labeled as "whites," "blacks," "reds," or "yellows." The process of abstraction that combined (a) the geographical distribution of a population, (b) its supposed skin color, and (c) its supposed constitution, was a feature distinctive of the work of certain scholars of the late seventeenth and early eighteenth century, and it was part of a more ambitious endeavor to classify reality. As far as mankind was concerned, this process took paradigmatic form in the classification drawn up by Linnaeus in the tenth edition of his *Systema naturæ* (1758–1759), where the varieties of the human species were classified as:

Americanus. rufus, cholericus, rectus.
Europæus. albus, sanguineus, torosus.

Asiaticus. luridus, melancholicus, rigidus.
Afer. niger, phlegmaticus, laxus.[19]

Each of the four terms designating human variety, indicates its geographical distribution, skin color or complexion, temperament and physique. The underlying scheme is the ancient quaternary system that Linnaeus introduced to this area of inquiry, and which was used for similar purposes by Kant in 1775.[20] The notion of species of men, varieties of men or races of men emerged mainly from the combination of skin color with geographical distribution, conceived within the ancient quaternary scheme of the human constitution. However, the core of this scheme was the black/white polarity, while the skin colors "red"[21] and "yellow"[22] were arbitrary abstractions based on selected sources along with the assignment of a typical temperament and physique to each human variety.

From a historical point of view, it should be stressed that the idea that each human variety could somehow be identified by skin color became an integral part of all subsequent classifications of the late eighteenth century and the first half of the nineteenth, like those devised by Blumenbach, Virey, Cuvier, Lawrence, Prichard, Bory de Saint-Vincent, and Broc.[23] This should be emphasized because the idea persisted among specialists, although some scholars argued at the end of the eighteenth century that skin color on its own was not enough to signal the differences among men. In 1787, for instance, while enumerating some of the theories concerning the causes that might produce black skin, Jefferson wrote:

> The first difference which strikes us is that of colour. Whether the black of the negro resides in the reticular membrane between the skin and scarf-skin itself; whether it proceeds from the colour of the blood, the colour of the bile, or from that of some other secretion, the difference is fixed in nature, and is real as if its seat and cause were better known to us. And is this difference of no importance? Is it not the foundation of a greater or less share of beauty in the two races?[24]

Aesthetic and intellectual features were the main criteria by which Jefferson distinguished blacks from whites:

> Comparing them [blacks] by their faculties of memory, reason, and imagination, it appears to me, that in memory they are equal to the whites; in reason much inferior, as I think one could scarcely be

found capable of tracing and comprehending the investigations of Euclid; and in imagination they are dull, tasteless, and anomalous.[25]

Neither the astronomical almanacs produced by the mathematical practitioner Benjamin Banneker, a free black who also participated in the survey of the Federal Territory (now the District of Columbia), nor the many slave narratives published by blacks during the abolitionist campaigns, ever changed Jefferson's opinion that no black could achieve anything intellectually significant without the help of a white person.[26]

The focus thus shifted from color to mental faculties and to analysis of the cranium. But, despite this shift, the main object of inquiry continued to be the polarity between Europeans and Africans; hence, comparisons were made between the brain volume of whites and blacks, or between their facial angles. The new classifications besides color, based on Pieter Camper's measurement of the facial angle and the shape of the cranium, yielded a different image of humanity from those exclusively based on color. While in the latter, humanity could be envisaged as a set of colors which merged into each other without being necessarily arranged in a hierarchical order (as in Buffon, Blumenbach, and Herder), the alignment of human and animal crania according to the increasing order of the facial angle, generated instead a sense of a hierarchical progression in the scale of beings.[27]

SKIN COLOR, SLAVERY, AND THE NOTION OF RACE

What, therefore, is the upshot of all this? From a historical point of view, it is that the notion of biological race came chronologically later than classifications of the human species based on skin color alone, and also that the debates relative to human pigmentation were the first categorization of physical difference. Color preceded the notion of race. This means, furthermore, that prejudice against colored peoples was chronologically antecedent to the idea that there existed human types or races, for which it provided the constitutive matrix. Any contemporary discussion that seeks to understand the roots of European racism should concern itself with those early debates, because the interpretative models forged by them persisted much longer and more pervasively than the notion of race, developing into the macro-models and stereotypes that still shape our language and behavior, and of which we are often unaware.

I have already pointed out that the first physical difference to become an effective cognitive problem for Europeans was the skin color of a population that it had known, at least in the Mediterranean area, since time immemorial. I have also pointed out the obsession with which Europeans studied the skin of Africans; I have mentioned their theological and physiological speculations on the origins of blackness, emphasizing that the first classifications of the varieties of the human species were based on the polarity between white and black. I now briefly discuss some of the motives underlying that interest and the accompanying studies on Africans and—as far as possible—show that they make some sense.

When the scholars of the late seventeenth century asked "Why are Ethiopians black?" they were certainly not immune to collective perceptions of Africans. These perceptions had been recast in the early sixteenth century when, for the first time in the history of mankind, the Europeans introduced a system of color-based slavery (or what is generally called *racial slavery*), which profoundly altered their perceptive evaluation of the peoples they subjugated. Color-prejudice, in fact, either did not exist or was very mild in the Greco-Roman Mediterranean world, and scant traces of it are to be found in the Middle Ages and the fifteenth century.[28] It gathered strength, however, in the course of the sixteenth century and was one of the consequences of color-based slavery.

In the mid-seventeenth century, the order that Europeans conferred on the colonies ranked the African as a slave, and for the majority of Christians, he was certainly a human person, though a descendant of a forebear bound in slavery. From a legal point of view, he was a commodity, and he was defined as such by the dictionaries of the time, while for some philosophers the color of his skin was so enigmatic that they raised doubts as to his human nature. The *literati* made use of the image of the sub-Saharan African according to the traditional symbolism of white and black, while some painters depicted him as a precious object included in a still life.

The conceptions just outlined can be borne out by a large quantity of sources, but what I wish to emphasize here is that the social and political classification of Africans preceded the scientific classifications and the anatomical investigations of skin color. When in the mid-seventeenth century, a European, even if a natural philosopher, saw a black man or the picture of one, he associated his skin color with a well-defined rank in the social hierarchy: a slave in the colonies, a valuable or prestigious commodity in Europe. This political-social classification presumably exerted a major influence on the studies and scientific classifications mentioned above. This does not imply, however, that those studies were motivated solely by an endeavor to give religious, rational, and natural philosophical legitimacy to slavery; otherwise one could not explain the many protests against slavery voiced by doctors and naturalists. In 1646, the English doctor Thomas Browne denounced two theories then current on the blackness of Africans as "vulgar errors"—what today would be called *prejudices*—and proposed a chemical explanation, which asserted that color was an integral part of nature.[29] The German-Dutch doctor Johannes Pechlin demonstrated the relativity of the concept of physical beauty and its inapplicability in science.[30] Moreover, in 1677, Pechlin began to emphasize that it had been the exploitation of Africans by Europeans, which modified the latter's perception of Africans! The works of Buffon, Camper, and Blumenbach contain passages in which they express sympathy for their Ethiopian "brothers" and condemn slavery.[31]

In 1765 Le Cat thought it was sheer madness to consider black skin the punishment of a crime.[32] One of the main results achieved by natural historians was the definition in the *Encyclopédie*: "NEGRE, f.m. (*Hist. Nat.*), homme qui habite différentes parties de la terre."[33] No previous French dictionary had included the word *homme* in its definition of "Negro." It was natural history, therefore, that showed that the sub-Saharan African was a human person and not a commodity. In 1808, after the abolition of the slave trade, in his book *Antropologia*, Thomas Jarrold said of Africans that "their persons are no longer merchandise. No longer merchandise!"[34] and Bory de Saint-Vincent, although a polygenist, included in his work of 1825 the most ferocious critique on slavery that I have found in a scientific work.[35] In his *Su I neri: saggio ideologico e fisiologico* of 1826, the Neapolitan Gaetano Pesce wrote that the races described by the naturalists were "devoid of reality" and "pure fictions of naturalists," while in the same year Abbé Grégoire, an illustrious abolitionist and scientist, wrote that "la différence de couleur est un accident physique qu'on a travesti en question politique."[36]

However, there was no lack of scholars, like Meiners, Virey, and Heusinger, who argued in favor of maintaining slavery and held that the category of "beauty" could act as a valid criterion of classification.[37] It is interesting to note what Virey had to say on the subject: "The negro is and always will be a slave; interest requires it, politics demands it, and his own constitution submits itself to it almost with no pain."[38] The idea that there were biologically superior and inferior human groups certainly also originated from the work of naturalists and more generally of scientists. When the first volume of Gobineau's *Essai sur l'inégalité des races humaines* came out in 1853, it was already a dominant doctrine. Nor did Gobineau's work appear in a vacuum.

It may be considered—and usually is—the beginning of racist ideology, but it should also be interpreted as the outcome of a long cultural and political process.[39]

In fact, from a chronological point of view, the idea gained ground and imposed itself as the majority view, just as the campaigns for abolition of first the slave trade and then slavery got under way, but especially after the revolts in the Caribbean and the enactment, on the July 8, 1801, of the first constitution of St. Domingue. Article 1 of the constitution stated that the colony was part of the French empire, but was ruled by "special laws." Article 3 established that slavery be abolished forever: "All its inhabitants are born, live and die free and French." Article 4 specified that any person, whatever "his color," was entitled to apply for all forms of employment, and Article 5 stated that no other distinction exists between men than that of virtues and talents and that all men are equal before the law, whether they are to be punished or protected.[40]

The principle of self-government (Art. 1.), the abolition of slavery (Art. 3.), the removal of all distinctions among whites, blacks, and mulattos, and therefore of the caste system (Art. 4.), and finally the statement of the principle of equality before the law with the consequent abolition of all the privileges, which hitherto had been enjoyed by the planters and suffered by the blacks and the mulattos (Art. 5.), represent the pillars of the 1801 Constitution of St. Domingue. Never before, and in no other colony, had blacks achieved anything similar: the accomplishment of a constitutional order about which not even the most enlightened European philosopher had ever dared to write. However, it was only through powerful structural social changes that such an extraordinary event came about; that is, through conflicts, revolts, revolutions, invasions, wars, and massacres on all sides, which had steeped St. Domingue in blood since July 1790, pitting "royalists against Republicans, masters against slaves, whites against colored, mulattos against blacks, invaders against invaded."[41] The revolts in St. Domingue caused a stir throughout Europe and changed the image that many scholars had of blacks. In his highly successful *Génie du christianisme*, published in April 1802, Chateaubriand wrote: "Who will now plead the cause of the blacks, after the crimes they have committed?"[42] The Danish naturalist Johann Christian Fabricius, a pupil of Linnaeus, well known for his studies of insects and for his reflections on hybridization and professor at the University of Kiel (then under Danish rule), published a book in 1804 in which he explicitly mentioned the events in St. Domingue. Apart from underlining the physical differences between whites and blacks—the color and structure of skin and hair, the form and bone structure of the skull, the prominence of the jaws, the thickness of the lips, the flat bone of the nose, the length of the forearm—Fabricius held that blacks lacked "acumen of intellect" because none of them had ever made a true discovery.[43] The absence of reflection, he wrote, is the reason why they had never liberated themselves from "the yoke of the whites."[44] "We annually lead them," he added, "as a flock of sheep from the coasts of Guinea to the West Indies," where we hold them by the thousands in slavery, and their frequent revolts against the whites have never been successful.[45] Should the blacks prevail over the French in St. Domingue, this would not depend on their capacity "to programme great plans with precision," but only on their number, the features of the region, the inhospitable climate, diseases, and the lack of French support troops.[46] Deploying the concept of an "intermediate fertile species" (*Mittelart*), Fabricius wrote, "For this reason I consider the black just as my half brother, generated by the cross of the white man and the ape."[47] The expression "half brother" was an implicit criticism against the English abolitionists' motto, "Am I not a Man and a Brother?"

ON THE CONSTRUCTION OF A EUROPEAN SOMATIC IDENTITY

I believe that it is impossible to get at the underlying sense of why Europeans considered color to be a mark of inferiority, thereby producing the stereotypes that have been transmitted in European culture to the present day, unless one asks the question: Why did the Europeans consider themselves to be superior? A clue is provided by certain definitions of "European" that are found in dictionaries and encyclopedias published during the period in question and in the views of eminent historians of the early nineteenth century.

The entry "European" appears rarely in the dictionaries and encyclopedias of the sixteenth and seventeenth centuries, while "Europe" is defined as the third, and then the fourth, part of the world. There are no entries on the subject in, for example, Calepino (1569), Florio (1659), Howell (1660), Richelet (1680–1688), Menagio (1685), Furetiere (1690), the *Vocabolario degli Accademici della Crusca* (1691), *Le dictionnaire de l'Académie françoise* (1694), Phillips (1706), and Richelet (1719). In the *Dictionnaire de Trévoux* of 1721 an entry states:

> Les *Européans* sont fils de Japhet; car l'Europe fut peuplée après le déluge par les enfans de ce fils de Noé [. . .] Les *Européans* sont les peuples de la terre les plus policez, les plus civilisez & les mieux

faites. Ils surpassent tous ceux des autres parties du Monde dans les Sciences & les Arts, & principalement dans ceux qu'on nomme libéraux, dans le commèrce, dans la navigation, dans la guèrre, dans les vèrtus militaires & civiles. Ils sont plus vaillants, plus prudents, plus généreux, plus doux, plus sociables, plus humains [. . .].[48]

The entry "Europe" in *Le grand dictionnaire historique*, published in 1759 by Louis Moréri, declares that:

Les peuples de l'Europe, par leur adresse & par leur courage, se sont soumis ceux des autres parties du monde; leur esprit paroît dans leurs ouvrages, leur sagesse dans leur gouvernement, leur force dans les armes, leur conduite dans le commerce, & la magnificence dans leurs villes. L'Europe surpasse aussi en toutes choses les autres parties du monde, soit pour ses édifices saints & profanes, soit pour le génie différent des peuples qui l'habitent. Nous pouvons encore ajouter aux avantages de l'Europe, celui d'avoir le vicaire de J. C. en terre dans la personne des papes.[49]

The *Encyclopédie* (1751–1765) also lacks an entry on "European," but includes a short one on "Europe,"[50] which states that, although it is the smallest of the four parts of the world, it has achieved such a high level of power that history has almost nothing to compare with it. The entry "Europeans" in *The New and Complete American Encyclopædia* (1805–1811), however contains a significant novelty:

EUROPEANS, The inhabitants of Europe. They are all white; and incomparably more handsome than the Africans, and even than most of the Asiatics. The Europeans surpass both in arts and sciences, especially in those called *liberal*; in trade, navigation, and in military and civil affairs; being at the same time, more prudent, more valiant, more generous, more polite, and more sociable than they: and though divided into various sects, yet as Christians, they have infinitely the advantage over a very large part of mankind. There are few places in Europe where men sell each other for slaves; and none where robbery is a profession, as it is in Asia and Africa.[51]

It will have been noted that this latter entry is almost a translation of the one in the *Dictionnaire de Trévoux*, but with a significant difference. The

Europeans, it is said, "are all white," a feature rarely to be found in any previous dictionary or encyclopedia.[52] And they are also contrasted with Africans and Asians from an aesthetic point of view.

The underlying sense, it seems to me, was that of a European somatic identity, in which whiteness was equated with freedom and blackness with slavery.[53] But it was a somatic identity mainly constructed on political-social relationships, and in particular on relationships of dominance and subjugation, which generated an ideology I have called *leucocracy*.[54] The otherness of the person destined for subjugation and exclusion—then and in the future—was explained in relation to the complexion of those who arrogated the right to subjugate and exclude. The claimed superiority of the "white man" was not dictated by nature; it arose from Europeans' evaluations of their own religion (Christianity) and military, commercial, and civil domination. *White* and *civilized* were terms so closely associated by early nineteenth-century historians, philosophers, and philologists, that they became almost synonymous. The same happened with the terms *colored* and *uncivilized*. Not surprisingly, therefore, the influential historian from Göttingen, Arnold Heeren, declared that the most difficult phenomenon for a historian to explain was the superiority of whites with respect to people of color:

> Whilst we see the surface of the other continents covered with nations of different, and almost always of dark color, (and, in so far as this determines the race, of different races,) the inhabitants of Europe belong only to one race. It has not, and it never had, any other native inhabitants than white nations. Is the white man distinguished by greater natural talents? [. . .] And yet we must esteem it probable; and how much does this probability increase in strength, if we make inquiries of history? The great superiority which the white nations in all ages and parts of the world have possessed, is a matter of fact, which cannot be done away with by denials. It may be said, this was the consequence of external circumstances, which favoured them more. But has this always been so? And why has it been so? And, further, why did those darker nations, which rose above the savage state, attain only to a degree of culture of their own; a degree which was passed neither by the Egyptian nor by the Mongolian, neither by the Chinese nor the Hindoo? And among the coloured races, why did the black remain behind the brown and the yellow? If these observations cannot but make us inclined to attribute differences of capacity to the several branches of our race, they do not on

that account prove an absolute want of capacity in our darker fellow-men [in German *Brüder*], nor must they be urged as containing the whole explanation of European superiority.[55]

In 1823, the French Protestant historian, François Guizot, defined superiority as "a living and expanding force" (the myth of Japhet the expansionist), which acted in fulfillment of a mission, which it itself did not know.[56] A few years later, he discovered that this mission was established by Providence: "European civilization has entered, if we may so speak, into the plan of Providence; it progresses according to the intentions of God. This is the rational account of its superiority."[57] The Europeans were a mystery even to themselves! To clarify the mystery they oriented the sciences to select tools of analysis that, by highlighting differences, fostered the growth of a powerful myth in which the construction of a biologically superior somatic identity was an integral part. First color, and later a notion of race, used as a biopolitical concept, were the constitutive elements of a mythological identity and of epidermic differences in others.

NOTES

1. Georges-Louis Leclerc, Comte de Buffon, "Histoire naturelle de l'homme," in *Histoire naturelle, générale et particuliére, avec la description du Cabinet du Roi* (Paris: de l'Imprimerie Royale, 1749), 3:305–530; Buffon, "De la dégénération des animaux," in *Histoire naturelle, générale et particuliére, avec la description du Cabinet du Roi* (Paris: de l'Imprimerie Royale, 1766), 14:311–374; Buffon, "Addition à l'article qui a pour titre, Variétés dans l'espèce humaine," in *Histoire naturelle, générale et particuliére, avec la description du Cabinet du Roi*, supplèment (Paris: de l'Imprimerie Royale, 1777), 4:454–582. On Buffon's anthropology and his sources, see especially Michèle Duchet, *Anthropologie et Histoire au siècle des lumières* (Paris: François Maspero, 1971), 229–280; and Jacques Roger, *Buffon un philosophe au Jardin du Roi* (Paris: Fayard, 1989), 301–351.
2. Isaac La Peyrère, *Præ-Adamitæ [. . .]* (np: np, 1655); La Peyrère, *Systema theologicum, ex Præadamitarum hypothesi: Pars prima* (np: np, 1655).
3. Giuliano Gliozzi, *Adamo e il nuovo mondo. La nascita dell'antropologia come ideologia coloniale: dalle genealogie bibliche alle teorie razziali (1500–1700)* (Firenze: La Nuova Italia editrice, 1977), 514–621.
4. Conrad Gesner, *Historiae animalium [. . .]*, 4 vols. (Tiguri: apud Christ.

Froschoverum, 1551–1558), 1:970. Andrew Battel, "The Strange Adventures of Andrew Battel [. . .]," *Hakluytus postumus [. . .]*, ed. Samuel Purchas, 5 vols. (London: by Stansby, 1625–1626), 2:970–985, 982. Jacob de Bondt, "Historiae naturalis [. . .]," in *De Indiae utriusque re naturali et medica [. . .]*, ed. Willem Piso (Amstelaedami: apud Ludovicum et Danielem Elzevirios, 1658), 50–86, 84–85. Edward Tyson, *Orang-utan, sive Homo sylvestris; or, The Anatomy of a Pygmie Compared with that of a Monkey, an Ape, and a Man [. . .]* (London: Thomas Bennet and Daniel Brown, 1699). See especially Giulio Barsanti, "Storia naturale delle scimmie 1600–1800," *Nuncius* 5, no. 2 (1990):99–165.

5. Richard Bradley, *A Philosophical Account of the Works of Nature, [. . .]* (London: printed for W. Mears, 1721), 95.

6. Carolus Linnaeus, *Systema naturae, sive regna tria naturae systematice proposita per classes, ordines, genera, & species [. . .]* (Lugduni Batavorum: apud Theodorum Haak, ex typographia Joannis Wilhelmi de Groot, 1735). On Linneaus' anthropology, see Gunnar Broberg, "Homo sapiens. Linnaeus's Classification of Man," in *Linnaeus: The Man and His Work*, ed. Tore Frängsmyr (Berkeley, Los Angeles, London: University of California Press, 1983), 156–194.

7. Letter from Linnaeus to Johann Georg Gmelin dated February 14, 1747, and published in *Joannis Georgii Gmelini, Reliquias quae supersunt commercii epistolici cum Carolo Linnaeo, Alberto Hallero, Guilielmo Stellero et al., [. . .], publicandas* curavit Guil. Henr. Theodor Plieninger (Stuttgartiae: typis C. F. Heringianis, 1861), 54–56. "quaero a Te et Toto orbe differentiam genericam inter hominem et Simiam, quae ex principiis Historiae naturalis. ego certissime nullam novi" (Ibid., 55).

8. Cornélius de Pauw, *Recherches philosophiques sur les Américains, ou Mémoires intéressants pour servir à l'histoire de l'espèce humaine* (Berlin: chez George Jacque Decker, 1768–69), 1:175–207.

9. Francesco Buzzi, "Dissertazione storico-anatomica sopra una varietà particolare d'Uomini bianchi Eliofobi," *Opuscoli scelti sulle scienze e sulle arti* 7 (1784):81–96; translated into German as: "Bemerkungen über die Kakerlaken," *Italienische medicinisch-chirurgische Bibliothek* 4/1 (1798):17–27. Johann Friedrich Blumenbach, "De oculis Leucaethiopum et iridis motu commentatio" [Read October 9, 1784], *Commentationes Societatis regiae scientiarum gottingensis ad a. MDCCLXXXIV et LXXXV*, 7 (1786):29–61; Blumenbach himself translated the first part of the essay into German as: "Von den Augen der Kakkerlaken," *Magazin für die Naturgeschichte des Menschen* 1/1 (1788), 71–88. For a history of

albinism, see Renato G. Mazzolini, "Albinos, Leucoæthiopes, Dondos, Kakerlakken: sulla storia dell'albinismo dal 1609 al 1812," in *La natura e il corpo: Studi in memoria di Attilio Zanca*, ed. Giuseppe Papagno and Giuseppe Olmi (Firenze: Leo S. Olschki, 2006), 161–204.

10. Renato G. Mazzolini, "*Las Castas*: Interracial Crossing and Social Structure, 1770–1835," in *Heredity Produced: At the Crossroads of Biology, Politics, and Culture, 1500–1870*, ed. Staffan Müller-Wille and Hans-Jörg Rheinberger (Cambridge, MA: MIT Press, 2007), 349–373. See also Staffan Müller-Wille, "Reproducing Difference: Race and Heredity from a *longue durée* Perspective" in this volume.

11. William Charles Wells, "An Account of a Female of the White Race of Mankind, Part of Whose Skin resembles that of a Negro; with some Observations on the Causes of the Differences in Colour and Form between the White and Negro Races of Men," in *Two Essays: One upon Single Vision with Two Eyes; the other on Dew [. . .]* (London: printed for Archibald Constable and Co., 1818), 423–439, especially, 432–436. For the relevance of this paper, which was recognized also by Charles Darwin, see Kentwood D. Wells, "William Charles Wells and the Races of Man," *Isis* 64 (1973):215–225.

12. The "mucous" or "reticular membrane" is now called *Malpighian* layer of the epidermis.

13. Marcello Malpighi, *De externo tactus organo anatomica observatio* (Neapoli: apud Ægidium Longu, 1665), 21–22. Johannes Nicolaas Pechlin, *De habitu & colore Æthiopum, qui vulgo nigritæ, liber* (Kiloni: literis ac impensis Joach. Reumanni, 1677). On Malpighi's contribution and Pechlin's dissection, see Renato G. Mazzolini, "Kiel 1675: la dissezione pubblica di una donna africana," in *Per una storia critica della scienza*, ed. Marco Beretta, Felice Mondella and Maria Teresa Monti (Milano: Cisalpino, 1996), 371–393.

14. Frederik Ruysch, *Johannis Gaubii Epistola problematica, prima ad [. . .] Fredericum Ruyschium [. . .] De pilis, pinguedine [. . .] de corpore reticulari, sub cuticula sito* (Amstelaedami: apud Johannes Wolters, 1696), table I, figs. 4, 5, 6, 7; Ruysch, *Thesaurus anatomicus primus.* (Amstelaedami: apud Joannem Wolters, 1701), table IV, figs. 8, 9; Id., *Thesaurus anatomicus secundus.* (Amstelaedami: apud Joannem Wolters, 1702), 62; Ruysch, *Thesaurus anatomicus tertius* (Amstelaedami: apud Joannem Wolters, 1703), 61; Ruysch, *Thesaurus anatomicus quintus.* (Amstelaedami: apud Joannem Wolters, 1705), 3–4; Ruysch, *Curae posteriores, seu thesaurus anatomicus omnium praecedentium maximus* (Amstelodami:

apud Janssonio-Waesbergios, 1724), 1–2; Bernhard Siegfried Albinus, *De sede et caussa coloris Æthiopum et cæterorum hominum* (Leidae Batavorum: apud Theodorum Haak; Amsterdam: apud Jacobum Graal & Henricum de Leth, 1737); Ruysch, *Academicarum annotationum*, 8 vols. (Leidae: apud J. & H. Verbeek, 1754–1768), Lib. I, table I, figs. 1, 2; Lib. III, table IV, figs. 1, 2, 3; Lib. VI, tables II and III. On the work of the two Dutch anatomists, see Renato G. Mazzolini, "Frammenti di pelle e immagini di uomini (1700–1740)," in *Natura-cultura. L'interpretazione del mondo fisico nei testi e nelle immagini*, ed. Giuseppe Olmi, Lucia Tongiorgi Tomasi, and Attilio Zanca (Firenze: Leo S. Olschki, 2000), 423–443.

15. Claude-Nicolas Le Cat, *Traité de la couleur de la peau humaine en général, de celle des nègres en particulier, et de la métamorphose d'une de ces couleurs en l'autre, soit de naissance, soit accidentellement* (Amsterdam: np, 1765), 57; Bartolomeo Bizio, "Ricerche chimiche sovra l'inchiostro della seppia," *Giornale di fisica, chimica, storia naturale, medicina, ed arti*, seconda decade, tomo 7 (1825), 88–108, p. 100; a German summary of this paper with the title, "Zur Zoochemie. Chemische Untersuchung der Sepientinte," appeared in *Journal für Chemie und Physik* 15 (1825):129–149, 135.

16. Johann Ludwig Hannemann, *Curiosum scrutinium nigredinis posterorum Cham i.e. Æthiopum: Juxta principia philosophiæ corpuscolaris adornatum* (Kiloni: literis & impensis Joach. Reumanni, 1677). On Noah's curse, see Norbert Klatt, *Verflucht, versklavt, verketzert: der verrußte Cham als Stammvater der Neger* (Göttingen: Klatt, 1998); Stephen R. Haynes, *Noah's Curse: The Biblical Justification of American Slavery* (Oxford, New York: Oxford University Press, 2002); David M. Goldenberg, *The Curse of Ham: Race and Slavery in Early Judaism, Christianity, and Islam* (Princeton and Oxford: Princeton University Press, 2003).

17. William Whiston, *A Supplement to the Literal Accomplishment of Scripture Prophecies. [. . .]* (London: Printed for J. Senex, 1725), 106–134. [Auguste Malfert], "Mémoire sur l'origine des Négres & des Américains," *Mémoires pour l'histoire des sciences & des beaux arts [Mémoires de Trévoux]* 65 (November, 1733):1927–1977.

18. William Frederick Van Amringe, *An Investigation of the Theories of the Natural History of Man, by Lawrence, Prichard, and others, founded upon Animal Analogies: And an Outline of a New Natural History of Man founded upon History, Anatomy, Physiology, and Human Analogies* (New York: Baker & Scribner, 1848):395–396.

19. Carolus Linnaeus, *Systema naturae [. . .]*, Editio decima reformata [. . .], 2 vols. (Holmiae: impensis L. Salvii, 1758–1759), 1:20–22. See also, *Systema naturae [. . .]*, Editio duodecima, reformata, 3 vols. in 4 parts (Holmiae: L. Salvii, 1766–68), 1:28–29; and, *Systema naturae [. . .]*, Editio decimo tertia, aucta, reformata. Cura Jo. Frif. Gmelin, 2 vols. (Lipsiae: impensis G. E. Beer, 1788), 1:21–23.
20. Immanuel Kant, *Von den verschiedenen Racen der Menschen zur Ankündigung der Vorlesungen der physischen Geographie im Sommerhalbenjahre 1775* (Königsberg: bey G. L. Hartung, 1775). This was also published, with significant changes and additions, as "Von den verschiedenen Rassen der Menschen," *Der Philosoph für die Welt*, ed. J. J. Engel, 22 (1777), 125–164. Kant, "Bestimmung des Begriffs einer Menschenrace," *Berlinische Monatsschrift* 6 (1785), 390–417. On Kant's concept of race, see Robert Bernasconi, "Who Invented the Concept of Race? Kant's Role in the Enlightenment Construction of Race." In *Race*, ed. Robert Bernasconi (Malden, MA and Oxford: Blackwell, 2001):11–36. See also Bernasconi's chapter in this volume "Heredity and Hybridity in the Natural History of Kant, Girtanner and Schelling during the 1790s."
21. Nancy Shoemaker, "How Indians Got to be Red," *American Historical Review* 102 (1997), 625–644. Shoemaker, *A Strange Likeness: Becoming Red and White in Eighteenth-Century North America* (Oxford: Oxford University Press, 2004).
22. Walter Demel, "Wie die Chinesen gelb wurden," *Historische Zeitschrift* 255 (1992), 625–666; Michael Keevak, *Becoming Yellow: A Short History of Racial Thinking* (Princeton and Oxford: Princeton University Press, 2011).
23. Johann Friedrich Blumenbach, *De generis humani varietate nativa [. . .]* (Goettingae: typis Frid. Andr. Rosenbuschii, 1775); Blumenbach, *Institutiones physiologicae* (Gottingae: apud Jo. Christ. Dieterich, 1787); Blumenbach, *De generis humani varietate nativa*, Editio tertia. (Gottingae, apud Vandenhoek et Ruprecht, 1795); Jules-Joseph Virey, *Histoire naturelle du genre humain, ou recherches sur ses principaux fondemens physiques et moraux, precedees d'un discours sur la nature des etres organiques, et sur l'ensemble de leur physiologie. On y a joint une dissertation sur le sauvage de l'Aveyron*, 2 vols. (Paris: chez Dufart, 1801); Virey, "Homme," in *Nouveau dictionnaire d'histoire naturelle [. . .]*, Nouvelle édition, (Paris: chez Deterville, 1817), 15:1–270; Virey, "Nègre," in *Nouveau dictionnaire d'histoire naturelle [. . .]*, Nouvelle édition (Paris: chez Deterville,

1818), 22:422–471; Virey, *Histoire naturelle du genre humain*, Nouvelle édition, augmentée et entierement refondue, avec figures, 3 vols. (Paris: chez Crochard, 1824); Georges Cuvier, *Le règne animal distribué d'après son organisation [. . .]* (Paris: chez Deterville, 1817), 1:81–100; William Lawrence, *Lectures on Physiology, Zoology, and the Natural History of Man [. . .]* (London: J. Callow, 1819). James Cowles Prichard, *Researches into the Physical History of Man* (London: John and Arthur Arch, 1813); Jean-Baptiste-Geneviève-Marcellin baron Bory de Saint-Vincent, "Homme," in *Dictionnaire classique d'histoire naturelle* (Paris: Rey et Gravier, 1825), 8:269–346; Saint-Vincent, *L'homme. (Homo.) Essai zoologique sur le genre humaine*, 2me édition, 2 vols. (Paris: Rey et Gravier, 1827); Pierre-Paul Broc, *Traité complet d'anatomie descriptive et raisonnée*, 3 vols. (Paris: Just Rouvier et E. Le Bouvier, 1833–1836); Saint-Vincent, *Des races humaines: Thèse pour le concours de la chaire d'anatomie, vacante à la Faculté de Médecine de Paris* (Paris: Imprimerie d'Hippolyte Tilliard, 1836); Saint-Vincent, *Essai sur les races humaines, considérées sous les rapports anatomiques et philosophique* (Bruxelles: Établissement encyclographique, 1837).

24. Thomas Jefferson, *Notes on the State of Virginia* (London: printed for John Stockdale, 1787), 229–230.
25. Jefferson, *Notes*, 232.
26. For Banneker's correspondence with Jefferson and the latter's judgment of Banneker, see Silvio a. Bedini, *The Life of Benjamin Banneker* (Rancho Cordova, CA: Landmark Enterprises, 1984), 152–158, 280–283. For a selection of slave narratives for the period 1770–1849, see volume one of *I Was Born a Slave: An Anthology of Classic Slave Narratives*, ed. Yuval Taylor, 2 vols. (Edinburgh: Charles Johnson, 1999).
27. Pieter (Petrus) Camper, *Verhandeling [. . .] over het natuurlijk verschil der wezenstrekken in menschen van onderscheiden landaart en ouderdom; over het schoon in antyke beelden en gesneedens steenen* (Utrecht: B. Wild en J. Altheer, 1791). This work was translated into French, German, and English. It should be noted that Camper was a staunch monogenist and did not associate the varieties of the human species with intelligence. As Miriam Claude Meijer has aptly shown, the facial angle theory took on a life of its own, which had nothing to do with Camper's general views on man, see Meijer, *Race and Aesthetics in the Anthropology of Petrus Camper (1722–1789)* (Amsterdam-Atlanta: Rodopi, 1999), 167–177. See also, Charles White, *An Account of the Regular Gradation in Man*

and in Different Animals and Vegetables; and from the former to the latter (London, printed for C. Dilly, 1799).
28. Frank M. Snowden, Jr., *Before Color Prejudice: The Ancient View of Blacks* (Cambridge, MA: Harvard University Press, 1983); Lellia Cracco Ruggini, "Il negro buono e il negro malvagio nel mondo classico," in *Conoscenze etniche e rapporti di convivenza nell'antichità*, ed. Marta Sordi (Milano: Vita e Pensiero, 1979), 108–135; Benjamin Braude, "Black Skin/White Skin in Ancient Greece and the Near East," *Micrologus* 13 (2005):11–21; Gude Suckale-Redlefsen, *Mauritius: Der heilige Mohr / The Black Saint Maurice* (Houston: Menil Foundation; München, Zürich, Verlag Schnell & Steiner, 1987); Maaike van der Lugt, "La peau noire dans la science médiévale," *Micrologus* 13 (2005):439–475.
29. Thomas Browne, *Pseudodoxia epidemica: Or, Enquiries into very many Received Tenents, and commonly Presumed Truths* (London: printed by T. H. for Edward Dod, 1646), 322–338. See also, Renato G. Mazzolini, "A Greater Division of Mankind is Made by the Skinne: Thomas Browne e il colore della pelle dei neri," *Micrologus* 13 (2005):571–604.
30. Pechlin, *De habitu & colore Æthiopum*, 105–106.
31. Buffon, "Histoire naturelle de l'homme," 305–530; Pieter (Petrus) Camper, "Redevoering over den oorsprong en de kleur der zwarten. Voogeleezen in den Ontleedkonstigen Schouwburg te Groningen, den 14 van slachtmaand 1764," *De rhapsodist* 2 (1772):373–394.
32. Le Cat, *Traité de la couleur de la peau*, 3.
33. *Encyclopédie, ou dictionnaire raisonné des sciences, des arts et des métiers, par une société de gens de lettres*, 17 vols. (Neufchastel, 1765), 11:76–83, 76: "Negro (*Natural History*), man inhabiting different parts of the earth."
34. Thomas Jarrold, *Anthropologia: Or, Dissertations on the Form and Colour of Man: with incidental remarks* (London: printed for Cadell and Davis, 1808), 259.
35. Bory de Saint-Vincent, "Homme," 269–346, pp. 318–320; Bory de Saint-Vincent, *L'homme. (Homo.) Essai zoologique sur le genre humain*, 3me édition, 2 vols. (Paris, Rey et Gravier, 1836), 2:51–62.
36. Gaetano Pesce, *Su i neri: saggio ideologico e fisiologico* (Napoli: presso Manfredi, 1926). "vuoti di realtà" (Ibid., 273); "pure finzioni de' naturalisti" (Ibid., 274). Henri Grégoire, *De la noblesse de la peau, ou du préjugé des blancs contre la couleur des africains et celle de leurs descendans noirs et sang-mêlés* (Paris: Baudouin Frères, 1826), 61: "color difference is a physical accident which one has disguised as a political issue."

37. Christoph Meiners, *Grundriß der Geschichte der Menschheit* (Lemgo: im Velage der Meyerschen Buchhandlung, 1785); Meiners, "Ueber die rechtmässigkeit des Negern-Handels," *Göttingisches historisches Magazin* 2 (1788):398–416; Meiners, "Ueber die Natur der Afrikanischen Neger, und die davon abhangende Befreyung, oder Einschränkung der Schwarzen," *Göttingisches historisches Magazin* 6 (1790):385–456; Meiners, "Von den Varietäten und Abarten der Neger," *Göttingisches historisches Magazin* 6 (1790):625–645; Meiners, "Ueber die Farben, und Schattierungen verschiedener Völker," *Neues Göttingisches historisches Magazin* 1/4 (1792):611–672. Karl Friedrich Heusinger, "Remarques sur la sécrétion du pigment noir, et la formation des poils," *Journal complémentaire du dictionaire des sciences médicales* 14 (1822):229–241; Meiners, *Untersuchungen über die anomale Kohlen- und Pigment-Bildung in dem menschlichen Körper* (Eisenach: bey Johann Friedrich Baereche, 1823); Meiners, "Noch einige Beiträge zur Lehre von der Absonderung der Pigment im thierischen Körper," *Deutsches Archiv für die Physiologie*, 8 (1823):37–44; Meiners, *Grundriss der physischen und psychischen Anthropologie für Aerzte und Nichtärzte* (Eisenach: bei Johann Friedrich Baerecke, 1829):70–71.

38. Jules-Joseph Virey, "Nègre," in *Nouveau dictionnaire d'histoire naturelle*, 24 vols. (Paris: Déterville, 1803–1804), 15:431–457. Quoted here from the Venetian edition, *Nouveau dictionnaire d'histoire naturelle* 31 vols. (Venise: de l'Imprimerie de Palese, 1804–1808), 15:486–518. "Le *nègre* est et sera toujours esclave; l'intérêt l'exige, la politique le demande, et sa constitution s'y soumet presque sans peine" (Ibid, 496).

39. Exemplified in the works of Victor Courtet de l'Isle, *La science politique fondée sur la science de l'homme, ou étude des races humaines sous le rapport philosophique, historique et social* (Paris: chez Bertrand, 1838); and Robert Knox, *The Races of Men* (1850, Mnemosyne reprint, 1969).

40. *Las constituciones de Haiti*, Recopilacion y estudio preliminar de Luis Mariñas Otero (Madrid: Ediciones Cultura Hispanica, 1968), 109–110.

41. Robin Blackburn, *The Overthrow of Colonial Slavery 1776–1848* (London, New York: Verso, 1988), 250. On the Haitian Revolution, see the classical study by C. L. R. James, *The Black Jacobins: Toussaint L'Ouverture and the San Domingo Revolution* (New York: Vintage, 1989) [1st ed., 1937]; and Nick Nesbitt, *Universal Emancipation: The Haitian Revolution and the Radical Enlightenment* (Charlottesville, London: University of Virginia Press, 2008).

42. "qui oserait encore plaider la cause des Noirs après les crimes qu'ils ont

commis?" François-Auguste [François-René de] Chateaubriand, *Génie du christianisme, ou Beautés de la religion chrétienne* 5 vols. (Paris: Chez Migneret, 1802), 4:189.
43. "Schärfe des Verstandes." Johann Christian Fabricius, *Resultate naturhistorischer Vorlesungen* (Kiel: in der neuen academischen Buchhandlung, 1804):213.
44. "von dem Joche der Weißen." Fabricius, *Resultate*, 213.
45. "Wir führen sie jährlich als eine Heerde Schaafe von der Küste Guinea nach Westindien." Fabricius, *Resultate*, 213
46. "große Pläne mit Genauigkeit zu entwerfen." Fabricius, *Resultate*, 214.
47. "Ich sehe deswegen auch den Schwarzen nur für meinen Halbbruder an, der aus einer Vermischung des weißen Menschen und des Affen antstanden." Fabricius, *Resultate*, 215.
48. *Dictionnaire universel François et Latin [. . .]*, 5 vols. (Paris: chez Florentin Delaulne, 1721), 2:1542 : "Europeans are the sons of Japhet; because Europe was peopled after the deluge by the sons of this son of Noah [. . .] Europeans are the most cultivated, civilized and handsome peoples of the earth. They surpass all those of the other parts of the world in sciences and arts, especially in those called liberal, in trade, navigation, in warfair, in military and civil virtues. They are more valiant, more prudent, more generous, more polite, more sociable, more humane [. . .]."
49. Louis Moréri, *Le grand dictionnaire historique [. . .]*, 2me édition, 10 vols. (Paris: chez les libraires associés, 1759), 4:315–316, 316: "By their dexterity and their courage the peoples of Europe have subdued those of the other parts of the world; their *esprit* is disclosed in their works, their wisdom in their governments, their strength in weapons, their conduct in trade, and their splendour in their towns. In everything else Europe also surpasses the other parts of the world, both for its secular and sacred buildings, and for the different genius of the peoples who live in it. We may also add to the advantages of Europe that of having the Vicar of Christ in the person of the popes."
50. *Encyclopédie, ou dictionnaire raisonné des sciences, des arts et des métiers, par une société de gens de lettres*, 17 vols. (Paris [from vol. 8 Neufchastel]: 1751–1765), 6:211–212.
51. *The New and Complete American Encyclopædia: or, Universal Dictionary of Arts and Sciences [. . .]*, 7 vols. (New York: printed and published by E. Low, 1805–1811), 3 (1807):417.
52. The statement, "The inhabitants [of Europe] are all white; and incomparably more handsome than the Africans, and even than most of the

Asiatics," may be found in the entry "Europe" in vol. 7 (1798), 39–40, 40 of the *Encyclopædia; or, a Dictionary of Arts, Sciences, and Miscellaneus Literature [. . .]* (Philadelphia: printed by Thomas Dobson), which lacks the entry "Europeans."

53. Nell Irvin Painter, *The History of White People* (New York, London: W. W. Norton and Company, 2010).
54. Renato G. Mazzolini, "Leucocrazia o dell'identità somatica degli europei," in *Identità collettive tra Medioevo ed età moderna*, ed. Paolo Prodi and Wolfgang Reinhard (Bologna: Clueb, 2002), 43–64. See also Sara Figal, "The Caucasian Slave Race: Beautiful Circassians and the Hybrid Origin of European Identity," in this volume.
55. Arnold Hermann Ludwig Heeren, *Ideen über Politik, den Verkehr und den Handel, der vornehmsten Völker der alten Welt*, 3 vols. (Göttingen: bey Vandenhoek und Ruprecht, 1793–1812), 3/1:6–7. English translation from *Ancient Greece*. Trans. George Bancroft. New and improved edition (London: Henry G. Bohn, 1847), ix.
56. François Guizot, *Essais sur l'histoire de France* (Paris: chez J. L. J. Brière, 1823), 62.
57. François Guizot, *Cours d'histoire moderne: Histoire générale de la civilisation en Europe, depuis la chute de l'empire romain jusqu'à la révolution française* (Paris: Pichon et Didier, éditeurs, 1828), 2ᵉ leçon:11–12. English translation from *The History of Civilization, from the Fall of the Roman Empire to the French Revolution*, trans. William Hazlitt, 3 vols. (London: David Bogue, 1846), 1:27.

7

THE CAUCASIAN SLAVE RACE

Beautiful Circassians and the
Hybrid Origin of European Identity

SARA FIGAL

Consider the following triangulation around a female figure:
1. Travel writers of the seventeenth and eighteenth century (unreliable, yet widely read) identify the most beautiful women in the world as the "primitive" Georgians and Circassians from the Caucasus mountains, additionally noting their high value on the Ottoman slave market;[1]
2. European fantasies of beautiful captives—often Circassians—languishing in a Sultan's harem inspire novels, plays, and operas during the eighteenth century, even while the fictions reinforce a disapproval of Ottoman "decadence" and the religion of Islam;[2]
3. In an era obsessed with origins, German scientists officially dub the "white" race of Europeans "Caucasian," citing the superior beauty of Georgian and Circassian women as proof that the origin of white Europeans—and of the human species overall—should be located in the Caucasus mountains.[3]

With these points as orientation, I would like to examine the figure of the fetishized female from the Caucasus, the Circassian or Georgian slave, who became an unlikely icon for racial theorists and their narratives of European superiority.[4] I focus particularly on those literary tropes of beauty and breeding that were translated by scientific discourse into empirical evidence for theories of human difference.

The "Caucasian" race was identified and named at the end of the eighteenth century by a German scientific community intent on systematically explaining human variation for the emerging fields of natural history, comparative anatomy, and physical anthropology. Debates about whether race was a "real" category existing in nature or a conceptual heuristic for scientific thinking circulated among a group of German philosophers and natural scientists that included Immanuel Kant, Johann Friedrich Blumenbach, Georg Forster, Christoph Meiners, and Johann Gottfried Herder. Scholars now generally credit the "invention of race" to this group, although Kant and Blumenbach are usually singled out as the most significant contributors.[5] It is true that, in 1795, Blumenbach settled on names for five primary races in the third edition of his *De generis humani varietate nativa* (On the natural variety of mankind), using the term *Caucasian* for the first time (along with *Mongolian, Ethiopian, American,* and *Malay*). Blumenbach also proposed that the Caucasian should be considered the "primeval" race, from which the other four had developed over time.[6] So why did he locate the origin of the species in the Caucasus mountains?

This question has been asked several times in recent decades, both in the context of the history of science, as with Stephen Jay Gould's revised edition of *The Mismeasure of Man* (1996), and in the context of whiteness studies, as with Bruce Baum's *The Rise and Fall of the Caucasian Race* (2006).[7] Both of these works recognize that the naming of the Caucasian race was a response to a tangle of ideas—for the most part unscientific—about beauty. As David Bindman has noted, Europeans believed that "the peoples beyond the eastern border of the Holy Roman Empire—the Circassians and others—were of exceptional contentment and beauty," a belief that contributed to the nineteenth-century understanding of an *Aryan* race.[8] While investigations of early race theories have pursued the long-term consequences of this theoretical shift that emphasized aesthetic superiority as "natural" evidence for explanations of racial differentiation, none to date has fully investigated what Gould calls the "mental machinery"—in this case, the cultural and literary catalysts—that produced the *Caucasian* nomenclature.

The tale told by late eighteenth-century scientists and philosophers of "white" racial identity emphasized a set of attributes that included European geography, white skin color, and both cultural and physical superiority. It is surprising, then, that the figure at the center of this constructed narrative of origins and hereditary transmission is the beautiful Circassian from the Caucasus mountains. European readers of travelogues and novels were familiar with this figure: Circassian and Georgian women in particular were celebrated as fair-skinned and dark-eyed, and they were fetishized as the prized possessions of Ottoman harems. Beyond lurid European fantasies of the seraglio and in the real world of human traffic, Circassian girls were regularly captured by traders—or sold by parents to traders—who knew they fetched the highest prices on the slave market in Constantinople.[9] While harem stories made for entertaining reading and postures of cross-cultural outrage, there was little to suggest an obvious reinscription of the Circassian beauty as the genealogical source of the European race.

From the vantage point of history, considering how racial theories were translated into arguments for European cultural and political hegemony throughout the world, this choice for the originary body of the Caucasian ideal seems preposterous. It locates the generatrix of Caucasian identity at home outside of Europe, beyond the bounds of the Holy Roman Empire. The Circassians and Georgians, those Caucasians with whom we shall be most concerned, lived outside of the community of "civilized" cultures; they were described by Europeans as wild and primitive.[10] Finally, they existed outside of orthodox Christendom and were identified by travel writers as animist heathens with vestiges of Christian and Muslim influence.[11] Their ambiguous otherness—their distance from all values ascribed to the superior European white Christian male—are hardly attributes that one would expect to find central to a narrative of European racial identity.

How, then, could such a conceptual move be possible? And not only possible, but acceptable to a scientific community heavily invested in European distinctiveness? Postcolonial theorists have advanced dynamic ideas of hybridity and creolization as a radical challenge to historically entrenched ideals of "authenticity," and, in this light particularly, it is intriguing to recognize that the eighteenth-century construction of the racial "authentic" is based at the outset on a circulating set of tales and references involving religious, linguistic, and biological amalgamation.

"THE HANDSOMEST WOMEN OF THE WORLD": BEAUTY AND DESIRE

Although the women of the Caucasus belonged to a culture that was decidedly non-European, their beauty was legendary, the stuff of myth and erotic fantasy. In Europe during the seventeenth, eighteenth, and even nineteenth centuries, these women—identified interchangeably and imprecisely as Georgians or Circassians—were celebrated in texts by German (and French and

English and Russian) writers, whether in travel literature, anthropological treatises, natural history texts, novels, plays, or philosophical essays, as the most beautiful human beings in the world.[12] Two works of nonfiction from the late seventeenth century illustrate what would become a customary perspective. The first of these is François Bernier's *Nouvelle division de la terre par les différentes espèces ou races qui l'habitent* (New Division of the Earth According to the Different Species or Races of Men, 1684), generally regarded as the first text in which the term *race* functioned as a dominant classification scheme for patterns of difference among human peoples. The text is an early witness to what would develop and circulate throughout Europe as the legendary figure of the "Circassian beauty":

> It cannot be said that the native and aboriginal women of Persia are beautiful, but this does not prevent the city of Isfahan from being filled with an infinity of very handsome women, as well as very handsome men, in consequence of the great number of handsome slaves who are brought there from Georgia and Circassia. The Turks have also a great number of very handsome women; . . . they have . . . an immense quantity of slaves who come to them from Mingrelia, Georgia, and Circassia, where, according to all the Levantines and all the travellers, the handsomest women of the world are to be found.[13]

Bernier's attention to degrees of beauty characterizing the world's various peoples suggests, without theorizing or analyzing, that an aesthetic evaluation of physiological difference is a meaningful element in the differentiation of human groups. The aesthetic quality—containing no small measure of erotic appeal, to gauge from his description of female beauties—finds its supreme example in the women of the Caucasus. In this, he is not alone. At about the same time, the great travel writer John Chardin described various peoples of the Caucasus, reporting this about the Georgians in particular:

> The Complexion of the Georgians is the most beautiful in all the East; and I can safely say, That I never saw an ill-favour'd Countenance in all that Country, either of the one or other Sex: but I have seen those that have had Angels Faces; Nature having bestow'd upon the Women of that Country Graces and Features, which are not other where to be seen: So that 'tis impossible to behold 'em without falling in Love.[14]

Chardin expands on his superlative evaluation of the Georgian women, finding something that transcends the ordinary in their "Angels Faces." This special gift of nature exerts a power on the beholder, whose observation—which, in the case of a travel writer like Chardin, aims to be dispassionate—is suddenly complicated by a "falling in Love." Lest we attribute this "love" to a pious respect for the angelic, it is well to read on, as Chardin continues that these overwhelming beauties "have an extraordinary addiction to the male sex" that contributes to the "torrent of uncleanness" marking their culture.[15] His fascination wavers between two extremes of female characterization: the Georgian woman is at once angelic and licentious, inspiring love and demanding sexual gratification. She is utterly irresistible to all who would gaze on her, wielding a power that doesn't require actual presence but can be felt even when mediated by Chardin's text. For there is no doubt that readers who were able to "behold 'em" only through Chardin's descriptions were entranced, and they paid their homage by extending textual tribute in citations over the next centuries.

Chardin's lingering note on the potent seduction of Georgian beauty, relatively contained in its tone, is repeatedly cited by later race theorists as authoritative justification of these women's preeminence in human racial history, producing the "superior" stock of Europeans while also refining various lesser lines with gracious generosity. When we read later citations of this seductive beauty, quoted in order to validate the iconic status of the Georgian/Circassian female for the Caucasian race, we are witnessing both the fate of a text and the invention of a biological legacy. For Chardin's text was distilled and blended with geographical, philosophical, popular, and natural scientific conceptual systems and writing traditions in order to produce a normative, scientific discourse of heritable racial traits.

In all likelihood, Chardin's text would not have found the resonance it did—at least in regard to theories of optimal breeding lines in the Caucasus—were it not for myriad other writers who echoed his judgment. And it is not only in Chardin's text that observations on Circassian and Georgian beauty wax to what becomes at times significant digression. Other authors, too, seem to get carried away, distracted by their own voluptuous descriptions. As an example of the endurance of the legend and the ongoing interruption of erotic distraction in an otherwise sober chronicle of travels in faraway lands, we might look to a text written in 1793 by the officer, book dealer, and sometimes-writer Hermann Henrichs. As the title makes clear, Henrichs's *Kurze Geschichte des Prinzen Heraclius, und des gegenwärtigen Zustandes von Georgien* (Brief history of Prince Heraclius and the current state of Georgia), presents itself as a discourse on conditions in Georgia that would be of use to the German reading public. His descriptions of the land, the peoples, and the customs he encounters are not, for the most part, positive: he sets his tone early in the book with a vehement condemnation of the regular sale of children by their parents to the Turks.[16] He also observes disparagingly that the Georgians are peculiar among nations insofar as they are not only ignorant, but proud of their ignorance. If we are to believe Henrichs, they are a population of drunken, impoverished heathens, worthy only of European scorn. When it comes to the women, however, Henrichs shifts the mood of his report dramatically and offers the following excursus:

> It would be irresponsible to pass over the Beauties of this land in silence, since it is on their account that one knows the name of Georgia. Who is unaware that a third of the Turkish seraglios are comprised of these lovable children [*diese liebenswürdige Kinder*]? And among these, how many fail to take the first place among the favorites of the oriental princes? And who can look at them, without being ravished by wonder?—Their form, facial features, complexion, and the regularity of their limbs constitute in nature what the creative genius of the artist shows us as an ideal.—What girl can show a more beautiful foot and thigh? Where does one see a more majestic gait and a more slender build? An arm so full and round, and such a beautiful hand? Long black hair falls against a blindingly white neck, and its beauty alone would be sufficient to drive a fiery youth to distraction before the large, fiery, black eyes enslaved him utterly.[17]

Henrichs, who has seen these "Beauties" himself, assures us that to see them is to be ravished. Like Chardin, his narration attests to a loss of control in the viewing moment. As observer *cum* writer, he is compelled in the end to describe not merely the beauty he has seen, but the effects of this beauty on himself. In an encounter that otherwise should be read as an "orientalizing" account of a European traveler through eastern lands, with all of the power-imbalance weighted in his favor, Henrichs acknowledges the viewer's passive status. In an encounter with a woman in the Caucasus, a (European) traveler has no choice but to fall in love; he has no control over himself at the moment of looking and so is ravished by wonder. The Georgian, in this case, enslaves the man who desires her.

The reader of such texts is not present to experience the sight of such beauty. Instead, his wonder, if it is evoked, can be only the result of the reading experience. Henrichs's text seems designed to catalyze a response that is more than simple understanding. Had he been content with limiting his remarks to an objective statement of praise or reference to the universal acclaim described in the first sentence of the cited passage, his praise would have been appropriate enough for his choice of genre—the informative report. Instead, Henrichs follows his gesture of acknowledgment toward the legendary "Beauties" with uncharacteristic textual excess; unable to mark the full stop of his own impressions, he makes a syntactic shift and extends his thought with a series of particulars marked by em dash separations. These extensions, or suggestions of something *more*, operate as a telescoping interpolation. Each em dash initiates a tighter focus of the textual description on the women, moving from their general appeal, to the variety of attributes that are praiseworthy (form, facial features, complexion, etc.), to particular body parts (foot, thigh, arm, hand) and finally to the specifics of hair (long, black, falling down the neck) and eyes (large, fiery, black) that are described in terms of the effect they have on men: they overwhelm and enslave. This rhetorical progression is a technique often employed by pornographic writing (coming into its own as a genre in the late eighteenth century), which typically moves from a general scenario to the focus on single (and interchangeable) body parts, and culminates in a description of the experience of succumbing to (orgasmic) pleasure.

EROTIC COMMODITIES OF TRADE: THE CIRCULATION OF CIRCASSIAN WOMEN

The erotic effect of "beauty" attributed to the women of the Caucasus is clearly a titillating pleasure for Henrichs, as it is for Chardin and Bernier. Such descriptions certainly left their mark on subsequent readers, who transmitted and amplified the legendary promises of pleasure through centuries of echo and citation. We do well, however, to remember that these women were not merely objects of fantasy or erotic commodities in a metaphorical sense. Circassian females were, in fact, the most highly prized and highly regulated objects of the Ottoman slave trade. Circassians were acquired regularly through raids, through tributes paid to the Ottomans, and through the regular sale of children by parents. This was well known to eighteenth-century Europeans: in a discussion of inoculation practices in East and West, Voltaire discussed the Circassian identification of daughters as a "chief commodity of trade" quite forthrightly, remarking that "they supply the beauties for the harems of the Grand Pasha, the Grand Sophi of Persia, and of those wealthy enough to buy and maintain such precious merchandise."[18] And, while François Bernier noted one hundred years earlier that "Christians and Jews are not allowed to buy a Circassian slave at Constantinople. They are reserved for the Turks alone,"[19] the acquisition of Circassian slaves by traveling Europeans was by no means unheard of in the eighteenth century. An example, famous in her day, was Charlotte Aïssé (1694–1733), a Circassian purchased at the age of four from a slave market by Charles de Ferriol, the French Ambassador to Constantinople. She is unusual insofar as she was raised in elevated social circles, provided inspiration for the Abbé Prévost's somewhat scandalous *roman à clef* entitled *Histoire d'une Grecque moderne* (History of a modern Greek, 1740), and functioned as an icon of virtue in the face of French decadence when her posthumous biography was published. An unassailable testimony to her cultural status is the interest shown in her by Voltaire, who edited her letters after her death. An anonymous biographical sketch was translated into German and published in 1809 in August von Kotzebue's journal *Die Biene* under the title, "Die kleine Sklavin, eine wahre Geschichte" (The little slave, a true story).[20] Mademoiselle Aïssé's virtue—and her point of fascination—lay in the fact that, while desired by all men (she was, after all, a Circassian), she resisted (almost) all of their attentions. Although her life was in no way typical, in one respect it reinforces a pattern: both biographical and fictionalized representations of Aïssé circle around her Circassian origins, her "rescue" from a harem fate, and her irresistible beauty with as much lurid interest as if she had become the seraglio concubine she once seemed destined to be.

Western fascination with the eroticized Circassian continued: in 1856, the *New York Daily Times* printed an article (reprinted from the *London Post*) reporting a glut on the slave market in Constantinople as a consequence of the Russian conquest of parts of the Caucasus. "Formerly a Circassian slave girl was pretty sure of being bought into a good family, where not only good treatment, but often rank and fortune awaited her; but at present low rates she may be taken by any huxter who never thought of keeping a slave before."[21] The reputation of Circassian beauty, conjoined with the desire to purchase and profit from it, prompted P. T. Barnum's attempt to buy, via proxy, a Circassian slave girl from Constantinople for his American Museum of curiosities.[22] After his attempt failed, he instead employed local talent with an exotic pseudonym, promoting her as "the purest example of the white race." Barnum's conversion of the legendary "Circassian Beauty" into an exhibition freak for U.S. urban viewers is the subject of Linda Frost's analysis, which focuses on the complex layering of racial ideas and gender norms during a period of social upheaval in the United States.[23] For our current interest, it suffices to emphasize that, in each of these various instances across more than a century, the Circassian woman functions as a signifier for an eroticized exoticism: one that could be purchased and circulated, and one that had currency in the Parisian salons and Göttingen scientific societies as well as in the sensational entertainment venues of lower Manhattan, across nations and languages and time.[24]

It does not take much analysis to place the eroticized harem beauty within well-known critiques of Western "Orientalism," as the figure evoked notions of a decadent and despotic, if sensually appealing, East and affirming the West's position of moral superiority. How, then—to return to the initial question of this essay—could such a figure intersect meaningfully with the theoretical composition of the European (Caucasian) race? What are the cognitive conditions underlying European narratives of race that promote the Circassian woman as the symbolic mother of white European stock, when at the same time she is in reality a female harem slave, a converted Muslim, the mother of hybrid lines? This certainly runs counter to what Renato Mazzolini calls in this volume "a European somatic identity," noting correctly that—at least consciously—it was explicitly one "in which whiteness was equated with freedom and blackness with slavery."[25] In what ways do erotic fantasies and racial theories intersect in the European imagination of hereditary origins and self-fashioning?

CROSSBREEDINGS: IMPROVING BLOOD AND BEAUTY

Let us return to Chardin, who speculated on the consequences of Caucasian enticement and advanced a theory, which united physical beauty and (reproductive) sex. He did this in a description of a less-beautiful people, the Persians, by recounting their attempt to improve their physical condition through crossbreeding. "The Persian Blood," Chardin explains, "is naturally thick; it may be seen in the Goebres, who are the remainder of the ancient Persians; they are homely, ill shap'd, dull, and have a rough Skin, and an Olive Complexion."[26] The presumption that visible traits are a manifestation (at least in part) of a particular quality of blood—in this case, "naturally thick"—is an expression of common belief rather than the product of scientific demonstration. As an entrenched folk belief, the blood as a repository of traits as well as their vehicle of transmission required neither explanation nor defense.

It is likewise both unproblematic and self-evident to Chardin that a mixing of the blood of two distinct peoples produces a visible sign of that mixture on the body of the offspring. Thus, he maintains that while the original Persian stock is extremely ungainly, certain subgroups have improved their collective beauty through a century of consistent and deliberate interbreeding with Georgian or Circassian women:

> In the other Parts of the Kingdom, the Persian Blood is now grown clearer, by the mixture of the Georgian and Circassian Blood, which is certainly the People of the World, which Nature favours most, both upon the Account of the Shape and Complexion, and of the Boldness and Courage; they are likewise Sprightly, Courtly, and Amorous.[27]

Writing a century before scientists were prepared to attempt a systematic theory of hereditary traits and racial types, Chardin simply presumes as common knowledge that physical characteristics unique to a particular people are transmitted—and transmuted—through patterns of reproduction. There is nothing in his tone to suggest that he is proposing anything surprising when he observed that an "ugly" people may be made more lovely by a steady infusion of foreign "blood." One noteworthy detail in Chardin's account, nonetheless, is the valence of the term *clear*. While in subsequent race theory, descriptive words like *clear* and *pure* come to designate racial lines that are not "contaminated" or "muddied" or "mixed" with other types generally considered inferior, Chardin uses *clear* as a positive sign of the results of crossbreeding. We might associate clarification here with the removal of, rather than the preservation of, ethnic or "racial" particularities: clearer blood for Chardin manifests itself through bodies that, through a process of mixing and refining, are less ethnically singular and thus more beautiful.

At the same time, however, the Georgian and Circassian women remain a kind of standard, an ideal type toward which other groups may breed (shedding the marks of their thicker blood and uglier bodies), but which itself never seems to deviate from its aesthetic norms. Somehow, the Georgians and Circassians are unaffected by the mixing; they are donors only. The possibility of maintaining this imbalance must be "credited" to the slave trade, whereby innumerable girls were taken and circulated as breeding stock, while sufficient "originals" remained in their primitive isolation in the Caucasus to maintain a critical mass of suppliers.

The ways in which Chardin's hierarchy of clear blood and physical beauty intersects with social rank and power in the Persian world is revealing of a critical asymmetry in the gender dynamics of this theory of breeding. Chardin observes that the elite and powerful men of Persia are nearly all the product of a Persian father and a mother from the Caucasus:

> There is scarce a Gentleman in Persia, whose Mother is not a Georgian, or a Circassian Woman; to begin with the King, who commonly is a Georgian, or a Circassian by the Mother's side; and whereas, that mixture begun above a hundred Years ago, the Female kind is grown fairer, as well as the other, and the Persian Women are now very handsome, and very well shap'd, tho' they are still inferior to the Georgians: As to the Men, they are commonly Tall, Straight, Ruddy, Vigorous, have a good Air, and a pleasant Countenance.[28]

The beautiful women from the Caucasus appear by such account to be little more than a prized breeding stock selected to improve the appearance of the powerful male lines in Persia. Indeed, Chardin declares, "had it not been for the Alliance before mention'd, the Nobility of Persia had been the ugliest Men in the World." Without Circassian mothers, they would have been ugly, but their Persian fathers would still have guaranteed their status as the nobility of Persia. Beauty or lack thereof does not affect their status in the world. By contrast, the Georgian and Circassian women who possess a coveted beauty but nothing else have no social signature of their own; they are represented simply as a commodity for sexual pleasure and optimized reproduction.

In intriguing ways, the idealized view of Circassian beauty and the lurid fantasies of harem slave girls are buried in the foundations of European identity by physical anthropologists, racial theorists, traveling "ethnographers," and proponents of a "medical police" intent on bringing eighteenth-century dreams of perfectibility to bear on the human species. It is they who build on Chardin's shift in focus from erotic fantasy to sexual reproduction and who cite the breeding value of the Circassian beauty as a factor in the physical quality of various peoples.

INVENTING THE CAUCASIANS

A crucial part of this story involves the history of racial classification, which was initially codified around the end of the eighteenth century. European scientists posited a system of racial identity that generally explained the current state of each group as the product of two forces: a process of degeneration and separation from a (no longer extant) original stock, and a process of at least potential development—ideally, improvement—of both cultural and physical attributes, along paths distinctly different for each race.

It is important to keep in mind that the Caucasus Mountains were and still are a highly fraught dividing line between Asia and Europe: territories in Caucasia are variably considered to be in one or both continents. Throughout history, the region has been less a border than a blending zone between Christian Europe and Muslim Asia. Further, the Caucasus is and was one of the most linguistically and culturally diverse regions on earth. Jean Chardin remarked while traveling through Georgia, "You shall meet here in this Country with Armenians, Greeks, Jews, Turks, Persians, Indians, Tartars, Juscovites and Europeans; and the Armenians are so numerous, that they exceed the Georgians."[29] It would seem that, with the designation *Caucasian*, the origins of the European were not so European after all.

Historians of racialist thinking have devoted much energy and insight into the eighteenth and nineteenth-century Europeans' insistent differentiation of "self" from Africans, South Americans, and Native Americans, all variously described as slaves, heathens, and primitives. I would like now to contrast this with what would, on the surface, seem to be an opposite gesture—namely, the Europeans' identification with Caucasians, except that they are also described as slaves, heathens, and primitives. The original European, if we follow the logic of racialist thinking, was not actually European, but was instead an indistinct figure from the liminal zone of the Caucasus, with its Babel of languages, ethnic loyalties, and cultural practices. In a racial system predicated on the idea of gradual degeneration from a *Stamm* (stem), it was not the Europeans but the people of the Caucasus—with contested and shifting national boundaries, trapped between Russia and the Orient, traded throughout the Ottoman empire and beyond—who are theorized and written about as being "closest" to the *Urstamm* (original stem). According to the logic of racial taxonomy, which moved from "origin" to "most flawed" (e.g., the African and the American races), it was actually this unbounded, non-Western, non-European, non-Christian group of people who were identified as the superior human stock. It was not the Europeans per se but the people of the Caucasus who were identified in the eighteenth century as the source—a hybrid hereditary source—of the newly racialized European present.

Vanessa Agnew repeats what has come to be a truism when she asserts that, "for all eighteenth-century theorists race was an oppositional category:

the European was the yardstick against which others were compared and contrasted."[30] It is all the more startling, then, to realize that, in the conceptual soup that produced the racial schematization of the Caucasian, the European "yardstick" does not apply. Instead, we are forced to make sense of the fact that the Caucasian is one of the least stable—and least investigated—of the identity categories produced at the end of the eighteenth century. Rather than marking a fixed identity, *Caucasian* is better described by Homi Bhabha's metaphor of a "third space," a discursive condition in which signs and symbols of culture remain in flux, continually subject to translation.[31] This flux is evident from the initial location of European racial identity within this unstable territory. A tangle of justifications include biblical tradition, historical geography, and conclusions drawn from the physical beauty of the inhabitants themselves, especially the women, and their capacity to transmit this beauty through reproductive lines.

The label *Caucasian* is first used by Christoph Meiners, the professor of *Weltweisheit* at the University of Göttingen, for his separation of the human kind into two races—the Caucasian and Mongolian. For his *Grundriß der Geschichte der Menschheit* (Sketch for a history of mankind, 1786) historical narratives drawn from the Bible, classical mythology, and folklore sufficed as valid grounds for his division.[32] Meiners drew on a conceptual residue of old beliefs in the population of the earth by Noah's three sons. While many mocked a persistent tendency to read the Bible historically, the idea of the Ark's landing in the Caucasus was still generally accepted as valid. The region of the Caucasus was further enriched with its mythological history: Prometheus had been chained to a rock in the Caucasus; Jason found the Golden Fleece and met Medea in the Caucasus; and Greek culture generally pointed to the region's central role in early human history.[33]

Meiners found that, at this phase of his theorizing, he only needed two primary categories—the Caucasian and the Mongolian. By the time he published a second, "very much improved" edition of the *Sketch* in 1793, he introduced his revisions with the announcement of new racial designations. Instead of geographically oriented terms, Meiners settled on descriptive names. Thus the race he had originally identified as Caucasian should be known simply as the "white and beautiful," and the Mongolian was renamed "dark-skinned and ugly."[34] Meiners did proceed in voluminous later writings to subdivide his races, a task that included the division of Europeans into an elaborate hierarchy, elevating the German "nation" over "ugly, effeminate Latin races."[35] But as long as Meiners treated the "beautiful" Europeans as a collective, he justified their identification with the Caucasus thus: "Almost all of the Sagas, and the history of the most ancient peoples indicate that the Caucasus and the planes that stretch south from the Caucasus are the cradle of at least half of the human species."[36]

Beyond myth and legend, however, the location of racial origins in the Caucasus was justified by the beauty of body transmitted through generations by its inhabitants. Meiners noted further on the Caucasian race: "This line is no longer entirely pure and unmixed in the Caucasus; nonetheless, the Caucasian peoples, and particularly the women, are the most beautiful in all of Asia."[37] In his first edition, Meiners declared they were the most beautiful in all the world; by 1793, he was ready to separate the white race into Asiatic and Celtic subdivisions.[38] Nonetheless, Meiners repeats both of these points—that the Caucasians, and particularly the Circassians and Georgians, produced supremely beautiful women, *and* that the Caucasian peoples were no longer as "pure" as they once were—in various writings over many years. In 1788, he translated and printed an anonymous French article for his own *Göttingisches historisches Magazin* entitled "On the Peoples of the Caucasus," which stressed that, while the region was now such a mix of religions, customs, and languages that the particular origin (*Ursprung*) of each people was impossible to distinguish, "the Georgians alone are an exception: they have remained unmixed (*unvermischt*), and we thus know them still today as a unique people." This exemplary—"unmixed"—people is celebrated further as independent, the leaders are regarded as regional protectors, and the Georgians—while themselves unmixed—are the *Volks-Stämme* or the genealogical source of their neighbors.[39] In yet another essay, he declares unequivocally that "the blood of the Georgians is the most beautiful in the Orient, and I might well say in the entire world."[40] The Circassians, noted Meiners in another essay, are distinguishable by such a blossoming beauty that one must ask why such a blessed people never produced "a genuinely enlightened nation" on their own soil.[41]

Meiners's ambivalent perspective reveals a gradual shift in his thinking that redirected European racial associations away from the Caucasus and toward blonde and blue-eyed Celts—a shift embraced and developed by later nineteenth-century raciologists. William Ripley's Lowell lectures at Columbia University in 1896, reassembled into book form in 1899, are heir to this legacy:

> Byzantine harem tales of Circassian beauty have not failed to influence opinion upon the subject of European origins. Not even the charm of mystery remains in support of a Caucasian race theory

today. In the present state of our knowledge, it is therefore difficult to excuse the statement of a recent authority, who still persists in the title *Homo Caucasicus* as applied to the peoples of Europe. It is not true that any of these Caucasians are even "somewhat typical." As a fact, they could never be typical of anything. The name covers nearly every physical type and family of language of the Eur-Asian continent, except, as we have said, that blond, tall, "Aryan"-speaking one to which the name has been specifically applied. It is all false; not only improbable, but absurd. The Caucasus is not a cradle—it is rather a grave—of peoples, of languages, of customs, and of physical types. Let us be assured of that point at the outset.[42]

The indignation with which Ripley attacked the implication of Caucasian origins for Europeans attests not only to the vicious tone of the "Aryanism" that dominated race discourse in the 1890s, but also to the resilience of the earlier theories of Circassian beauty and European lineage that were written into the earliest definitions of "race" as a scientific category.

Nineteenth-century attempts to disentangle the construction of the European from the Caucasus were not limited to arguments about physical type or cultural practice. Hegel, in his *Encyclopaedia of the Philosophical Sciences* of 1830, praised the capacity of the Caucasian race—and no other—to create world history, but stressed the following condition: "In this, however, we have to distinguish two sides, the Western Asiatics and the Europeans; this distinction now coincides with that of Mohammedans and Christians."[43] The gesture here occludes the hybrid origin of the Caucasian, in part by oversimplifying the peoples grouped within the racial designation as clearly either Christian or Muslim. The particulars that comprise the original argument for "Caucasian" identity are retracted to the degree that the "Caucasian" no longer has ties to the person from the Caucasus; nevertheless, the term—with all of its echoes—remained intact.

But back to our defining moment—the late eighteenth century. Meiners's ambivalence notwithstanding, most of his contemporaries at the end of the eighteenth century continued to refer to Caucasian beauty as evidence of how deserving Caucasian women were of preeminence in human racial history, and they referred regularly to the ongoing value of what we might understand in eugenic terms as reproductive improvement of entire peoples when bred with beautiful Circassians. Although Meiners had introduced *Caucasian* as a term of racial classification in his early writings, the official designation of the white European race as *Caucasian* is historically ascribed to Blumenbach.

Blumenbach's first book, a widely read text on human difference entitled, *Über die natürlichen Verschiedenheiten im Menschengeschlechte* (On the natural variety of mankind), initially proposed four major races when it appeared in dissertation form in 1775; he adjusted the number to five by the time the book was reissued in 1781 and refined his nomenclature when he significantly revised the text for printing again in 1795.[44] Only in this version did he introduce the term *Caucasian*. *On the natural variety of mankind* reflects a contemporary reliance on "evidence," compiled from geographical and ethnocultural descriptions provided by the plethora of travel narratives published during the century, as skin color, hair texture, and skeletal (primarily cranial) difference; the later edition incorporates these observations with theories of natural teleological development. After decades of writing on the subject of human variety, why would Blumenbach adopt the term *Caucasian* in 1795? Skeptical of Meiners's methods and sweeping claims about non-European cultures, he did not cite Meiners as a source for this new terminology.[45] As Bruce Baum interprets the debt, Blumenbach "worked to distance his own anthropological thinking from that of Meiners while recovering the term *Caucasian* for his own more refined racial classification."[46] In his previous writings on racial divisions, Blumenbach used designations such as *European* and *White*. He justified his introduction of *Caucasian* to the nomenclature by pointing out that the area around Mount Caucasus "produces the most beautiful race of men, I mean the Georgian."[47]

This superior beauty, by sheer force of assertion and repetition, was thus rhetorically shifted from the realm of opinion and put to work as a fact that functioned as evidence and illustration of numerous conceptual offshoots of racial theorizing, including Caucasian identity. For example, when arguing that the "racial face" was, in instances of mixed-race breeding, a composite, Blumenbach offered as a typical example the blending of what he took to be extremes:

> These kinds of racial face, just like the colour of the skin, become mingled and as it were run together in the offspring from the unions of different varieties of mankind, so that the children present a countenance which is a mean between either parent. Hence . . . the offspring of the Nogay Tartars is rendered more beautiful through unions with the Georgians.[48]

Blumenbach's example underscores a conceptual arc of species-description that is anchored on one end by an extremity identified with ugliness and the Nogay Tartars and, on the other end, by an extremity identified with beauty and the Georgians. In each case, a particular people is named in order to clarify and embody the concept that Blumenbach is working out via his scientific theories and aesthetic evaluations, as if the ugliness of the Tartars and the beauty of the Georgians were indisputable and dispassionately cited facts rather than culturally relative judgments. Lavater, in his *Physiognomische Fragmente* (1775–1778), demonstrates the degree to which the Circassians (and the Tartars) functioned synechdocally when he said of the high degree of variation among African peoples that "they also have their Tartars and their Circassians," demarking thereby the dramatic range from ugliness to uncontested beauty.[49]

Blumenbach made a similar conceptual move when constructing evidence to argue that all human races belong to one species based on their ability to produce fertile offspring together. He wrote: "This definition of species may be conveniently illustrated [. . .]. Take . . . a man and a woman most widely different from each other; let the one be a most beautiful Circassian woman and the other an African born in Guinea, as black and ugly as possible [. . .]."[50] Their progeny, Blumenbach assures us, by definition, testifies to the species identity of both parents, and also demonstrates a blending of extreme traits. Beyond the species argument, which is Blumenbach's focus in this segment of text, his example betrays the fact that "ugliness" does not require a fixed point; ugliness, even at its extremity, may be exemplified by Nogay Tartars *or* by Africans from Guinea. However, the apex of human beauty remains the female of the Caucasus throughout his writings. Not only do these women embody an aesthetic ideal (which Blumenbach acknowledges is a construct of European bias), they are also valuable to the species insofar as such Circassian and Georgian females may be used to "improve" less beautiful peoples.

PROTO-EUGENICS, EXOTIC DESIRE, AND WHITE WOMEN

The identification of Circassians as exemplary breeding stock is present in a range of eighteenth-century texts that seized on one of the era's most persistent preoccupations, namely: the improvement of the human, be it as an individual or as a collective people, race, or species. In musing on the physical and moral improvement of human kind at mid-century, the French physician and hygienic theorist Charles Augustin Vandermonde attributed the cultural flourishing of great cities to their hybrid populations. Specifically, he praised

Turkey as one of the greatest states in the world and he focused on its "beautiful blood." This, Vandermonde stressed, was the admixture of many foreign peoples over many centuries, but he laid special emphasis on the "prodigious quantity of Georgian, Mingrelian, and Circassian slaves," who bequeathed their beauty to the Turkish stock.[51] The lessons rendered by such observations focused on the gains that would accrue to the human species if breeding practices were regulated and various human kinds mixed to optimal effect.[52]

Following Vandermonde's nascent "biopolitics," we find a more developed system of proto-eugenics in the work of Johann Peter Frank, the German physician, professor, and advocate of hygienic social policy. He pioneered the field of "medical police," a branch of *Polizeywissenschaft* that focused on issues of public health, and he contributed to the field the enormously influential, multivolume treatise, the *System einer vollständigen medicinischen Polizey* (A complete system of medical police, 1779–1827). Frank developed and advocated legally regulated breeding policies, for which he drew on available knowledge of hereditary transmission, theories of racial degeneration, and theories of disease transmission.[53] Although his eugenic program appears, by twenty-first-century standards, to be both oppressive and predictable in its curtailing of reproductive rights to people deemed congenitally "unhealthy" (including those displaying nonstandard form, stature, mental ability), Frank was remarkably unconventional in his presentation of intermarriage among different peoples or "races."[54] Unlike the majority of nineteenth-century thinkers who would follow him, Frank theorized interracial desire as both natural and as beneficial to the species. Interbreeding was beneficial insofar as it brought a "balance" to the color, stature, features, and general health of the progeny, and helped to rid the species of genetic weaknesses that resulted from inbreeding. Interracial desire was natural, Frank argued, simply because men desire "different" or "exotic" women. His examples focused on "mixed marriages" between Europeans (Caucasians) and Mongols. Frank states that European men wished to marry Mongol girls because, in his words, they are "hot-blooded." Conversely, wealthy Mongol men wished to marry Russian (Caucasian, Georgian) girls from poor families because of their superior beauty. In each case, women are identified as the object of male sexual desire, whereas men offer women increased social, racial, or financial status.

Frank theorizes this erotic desire specifically as a sign of an instinctive inclination toward hygienic health. In the example of European men marrying Mongol women, Frank identifies a "natural" desire for the racially exotic other. This attraction between different ethnic peoples results in strong and particularly beautiful progeny (provided that civic institutions and social customs allow such unions). There is at work here an anticipation of sexual selection theory, in the insinuation of a universal male sexual desire for a beauty in women that specifically balances a man's racial traits in the production of progeny. Frank's favored example returns us to the legendary Caucasian female—both to her uncontested beauty and to her slave status—as he reports how "the half-tartar Persian mitigates his natural ugliness through mixing his blood with that of the beautiful slave-girl from Teflis [Georgia]."[55]

This identification of a shrewd use of Georgian or Circassian stock by various "uglier" peoples tropes the female from the Caucasus as the ultimate medium for racial assimilation. In many such references to beautiful women and their capacity to improve other stock, we might identify a colonial fantasy carried out through reproductive assimilation. Women are offered up to the "ugly" (and inevitably darker) men of other races; in the case of the Ottomans (as with the seraglio fictions and comments on Persian stock), the other "race" may well be a military and cultural threat. In such a case, the improvement of the darker kind through Caucasian blood might well function as a biological conquering through racial "improvement." Logic would suggest that if everyone were to mate with a Caucasian woman, then all peoples would become more beautiful, more nearly European. This scenario, however, would require an endless supply of Georgian and Circassian women, an endless source of new origins, an endlessly pure maternal line.

The story of the justification for Caucasian as an identifying term for the European race is just that: a story told and retold through a plethora of textual variations. These variations produced and occluded multiple displacements. Felicity Nussbaum, discussing representations of female subjectivity in eighteenth-century Britain, argues that "women's empowerment in this period whether it derives from beauty's empire, linguistic skill, or political and military victory, is deeply bonded to defect and deformity."[56] If we accept this premise as holding generally true for European cultures suspicious of female accomplishment, we might well read an obverse proposition in the European idealization of the Circassian woman. Insofar as she is celebrated precisely as someone without land or language or religion of her own, either as a primitive or as a slave without personal freedom, she is the opposite of the "deformed" individualistic European woman. Instead, she embodies the site of two ideals simultaneously, the female (both erotic and reproductive) and the racial. Nussbaum has argued further that, at the end of the eighteenth century, "white women's sexuality becomes the carefully guarded line between the infected and uninfected spaces of racial and cultural contagion."[57] Her argument holds as long as the category of "white woman" is restricted to

Europe. When we consider how the woman from the Caucasus becomes the ideal white woman, whose body is physical proof of the European connection to a lost, original human ideal type, the story is far more complicated. In the case of the "Circassian beauty," the enslaved sexuality of the white woman (the embodied evidence of *Caucasian* identity) becomes the liminal space of contagion, the site of infection and transmission and purity all at once.

NOTES

1. Examples include: Jan Struys, *Drie aanmerkelyke en seer rampsoedige Reysen [. . .]* (Amsterdam: Van Meurs en van Someren, 1676), translated into English as *The perillous and most unhappy voyages of John Struys [. . .]* (London: S. Smith, 1683); François Bernier, "Nouvelle division de la terre par les différents espèces ou races qui l'habitent," *Journal des sçavans* April 24 (1684); John Chardin, *The Travels of Sir John Chardin into Persia and the East-Indies, Through the Black Sea, and the Country of Colchis* (London: np, 1691); George Sandys, *A Relation of a Journey begun An: DomL 1610. Foure Bookes Containing a description of the Turkish Empire [. . .]* (London: np, 1615), translated into German as *Sandys Reisen: Die Historie von dem ursprünglichen und gegenwertigem Stand deß Türkischen Reichs [. . .]* (Frankfurt am Main: np, 1696); Jakob Reineggs, *Allgemeine historisch-topographische Beschreibung des Kaukasus* (Gotha, St. Petersburg: np, 1796); Hermann Henrichs, *Kurze Geschichte des Prinzen Heraclius, und des gegenwärtigen Zustandes von Georgien* (Flensburg, Leipzig: np, 1793); Anonymous, "Der Kaukasus," *Konstantinopel und St. Petersburg, der Orient und der Norden*, (1805), 1:303–354.
2. Examples include Penelope Aubin, *The Noble Slaves* (1722); Voltaire, *Zaïre* (1732); Denis Diderot, *Les bijoux indiscrets* (1748); Christoph Martin Wieland, *Der goldene Spiegel* (1772); Wolfgang Amadeus Mozart and Christoph Friedrich Bretzner's *Die Entführung aus dem Serail* (1782). For a critique of harem accounts produced in Europe during the seventeenth and eighteenth centuries, see Alain Grosrichard, *The Sultan's Court: European Fantasies of the East*, trans. Liz Heron (London: Verso, 1998). For a discussion of the harem seduction/abduction motif in the eighteenth century, see also W. Daniel Wilson, *Humanität und Kreuzzugsideologie um 1780* (New York: Peter Lang, 1984), esp. 11–37.
3. For an overview of the development of scientific consensus, see Bruce Baum, *The Rise and Fall of the Caucasian Race: A Political History of Racial Identity* (New York, London: New York University Press, 2006).

4. During the eighteenth century, references to women of the Caucasus draw from travel reports of both Georgia and Circassia, such that the two regions become for all practical purposes indistinguishable for texts referring to exemplary beauty.
5. See, for example, Sara Eigen Figal, *Heredity, Race, and the Birth of the Modern* (London, New York: Routledge, 2008), esp. 59–84; Robert Bernasconi, "Who Invented the Concept of Race? Kant's Role in the Enlightenment Construction of Race," in *Race*, ed. Robert Bernasconi (Malden, MA, Oxford: Blackwell, 2001), 11–36; John H. Zammito, "Policing Polygeneticism in Germany, 1775: (Kames,) Kant and Blumenbach," in *The German Invention of Race*, ed. Sara Eigen and Mark Larrimore (Albany, NY: State University of New York Press, 2006), 35–54.
6. Johann Friedrich Blumenbach, *On the Natural Variety of Mankind*, in *The Anthropological Treatises of Johann Friedrich Blumenbach*, ed. and trans. Thomas Bendyshe (London, 1865), 264.
7. Stephen Jay Gould, *The Mismeasure of Man* (New York: Norton, 1996). Bruce Baum, *The Rise and Fall of the Caucasian Race*.
8. David Bindman, *Ape to Apollo: Aesthetics and the Idea of Race in the 18th Century* (Ithaca, NY: Cornell University Press, 2002), 25.
9. As William Clarence-Smith explains: "Circassians of the northwestern Caucasus, gradually embracing Islam from 1717, continued to sell servile soldiers and concubines after conversion, which was more or less complete around 1850. Some were constrained by poverty, but others wished to better the lot of their children. Their greatest hope was that a daughter might become 'the mother of sultans'." William Clarence-Smith, *Islam and the Abolition of Slavery* (Oxford: Oxford University Press, 2006), 74.
10. See, for example: Jakob Reineggs, "Kurzer Auszug der Geschichte von Georgien," *Neue Nordische Beyträge* 3 (1782):324.
11. See, for example, Adam Olearius, *Moskowitische und Persische Reise 1633–1639* (Stuttgart: Thienemann, 1986), 330; Denis Diderot, "Circassie," *Encyclopédie ou Dictionnaire raisonné des sciences, des arts et des métiers, par une Société de Gens de lettres* (1753), 3:458; G. Ellis, "Von den vornemsten Caucasischen Nationen," *Neue Beiträge zur Völker- und Länderkunde* 10 (1792):185.
12. Blumenbach cites John Chardin, the seventeenth-century travel writer, as only one of a "cloud of eye-witnesses." Blumenbach, *On the Natural Variety of Mankind*, 269, fn. 1; Baum, in *The Rise and Fall of the Caucasian Race*, cites a wide range of natural scientists (Bernier, Buffon, Goldsmith, Blumenbach, Meiners, Herder, Kant, Cuvier, Gobineau, etc.), who use beauty as evidence of "white" racial superiority. Other textual witnesses

to Georgian or Circassian beauty include Alexander Pope, Abbé Prévost, Charles Augustin Vandermonde, Henry Fielding, Christoph Martin Wieland, Johann Winckelmann, Peter Simon Pallas, Christian Lavater, and Alexander Pushkin. Baum, *The Rise and Fall of the Caucasian Race*. And let us not overlook the appearance on the popular front, with such works as *The Fair Circassian: A Dramatic Performance* published in London in 1720 and reprinted into the nineteenth century. This piece was initially issued anonymously—it is quite racy—by Samuel Croxall, the Archdeacon of Hereford, author of an enormously popular *Fables of Aesop* (London: np, 1722).

13. François Bernier, "A New Division of the Earth," in *The Idea of Race*, ed. Robert Bernasconi and Tommy L. Lott (Indianapolis: Hackett Publishing, 2000), 4.
14. Chardin, *The Travels of Sir John Chardin into Persia and the East-Indies, Through the Black Sea, and the Country of Colchis* (London: np, 1691), 190.
15. Chardin, *Travels*, 191.
16. Hermann Henrichs, *Kurze Geschichte des Prinzen Heraclius, und des gegenwärtigen Zustandes von Georgien* (Flensburg, Leipzig: Korten, 1793), 9.
17. Henrichs, *Kurze Geschichte*, 39–40.
18. François Marie Arouet de Voltaire, "On Inoculation with Smallpox," *Philosophical Letters, Or, Letters Regarding the English Nation*, trans. Prudence L. Steiner (Indianapolis: Hackett Publishing, 2007), 33.
19. Bernier, "A New Division of the Earth," 4.
20. Anonymous, "Die kleine Sklavin, eine wahre Geschichte," *Die Biene oder neue kleine Schriften* 3 (Königsberg: np, 1809):129–180.
21. *New York Daily Times*, August 6, 1856, 6.
22. See Linda Frost, "The Circassian Beauty and the Circassian Slave: Gender, Imperialism, and American Popular Entertainment," in *Freakery: Cultural Spectacles of the Extraordinary Body*, ed. Rosemarie Garland Thomson (New York, London: New York University Press, 1996), 248–262.
23. Linda Frost, *Never One Nation: Freaks, Savages, and Whiteness in U.S. Popular Culture, 1850–1877* (Minneapolis: University of Minnesota Press, 2005).
24. In October 2008, a website advertising Georgian tourism attracts English-speaking travelers with a heady list of citations culled from texts produced over nearly two millennia that all attest to the beauty of the population. See http://www.mcitours.ge/?m=publications&c=22.
25. Renato G. Mazzolini, "Skin Color and the Origin of Physical Anthropology (1640–1850)," in this volume.

26. Sir John Chardin, *Sir John Chardin's Travels in Persia. Never before translated into English [...]* (London: np, 1720), 2:119.
27. Chardin, *Travels in Persia.*
28. Chardin, *Sir John Chardin's Travels in Persia,* 119–120.
29. Chardin, *The Travels of Sir John Chardin,* 191.
30. Vanessa Agnew, "Pacific Island Encounters and the German Invention of Race," in *Islands in History and Representation,* ed. Rod Edmond and Vanessa Smith (London, New York: Routledge, 2003), 91.
31. Homi Bhabha, *The Location of Culture* (London, New York: Routledge, 1994), 37.
32. Christoph Meiners, *Grundriß der Geschichte der Menschheit* (Lemgo: Meyersche Buchhandlung, 1785).
33. As Nell Irvin Painter notes wryly, "Western Europeans had long traced their origins to amorphous Eurasian regions" but always located these ancestors in a clearly pre-Muslim antiquity. "Why Are White People Called 'Caucasian'?" *Proceedings of the Fifth Annual Gilder Lehrman Center International Conference at Yale University: "Collective Degradation: Slavery and the Construction of Race"* November (2003), 25, http://www.yale.edu/glc/events/race/Painter.pdf.
34. Christoph Meiners, *Grundriß der Geschichte der Menschheit: Zweyte sehr verbesserte Ausgabe* (Lemgo: Meyersche Buchhandlung, 1793).
35. Baum, *The Rise and Fall of the Caucasian Race,* 87.
36. Meiners, *Grundriß der Geschichte der Menschheit,* 47.
37. Ibid., 74.
38. Ibid., 75.
39. This is a text that Meiners translated from the French (author anonymous) and published in his magazine: "Ueber die Völkerschaften des Kaukasus," *Göttingisches historisches Magazin* (1788):110–111.
40. Christoph Meiners, "Historische Betrachtungen über die Fruchtbarkeit, oder Unfruchtbarkeit der Bevölkerung, oder Entvölkerung der verschiedenen Erdtheile," *Neues Göttingisches historisches Magazin* (1794):550.
41. Meiners, "Historische Betrachtungen," 555.
42. William Ripley, *The Races of Europe: A Sociological Study* (New York: D. Appleton and Co., 1899), 437.
43. Georg Friedrich Wilhelm Hegel, "Anthropology," from the *Encyclopaedia of the Philosophical Sciences,* in ed. Robert Bernasconi and Tommy L. Lott, *The Idea of Race,* op. cit., 42
44. Blumenbach's dissertation (presented in Latin as *De generis humani varietate nativa liber*) was first published as book in 1775, reissued in 1781 and again in 1795.

45. On the connections between Blumenbach and Meiners, see John Zammito, "Policing Polygeneticism in Germany," 1775: (Kames,) Kant and Blimenbach, in *The German Invention of Race*, ed. Sara Eigen and Mark Larrimore (Albany, NY: State University of New York Press, 2006), 35–54.
46. Baum, *The Rise and Fall of the Caucasian Race*, 88.
47. Johann Friedrich Blumenbach, *De generis humani varietate nativa liber*, in *The Anthropological Treatises of Johann Friedrich Blumenbach*, trans. and ed. Thomas Bendyshe (London: np, 1865), 269.
48. Blumenbach, *De generis humani*, 233. Blumenbach cites here M. de Peyssonel, *Traite sur le commerce de la mer Noire* (Paris: np, 1787).
49. Johann Caspar Lavater, *Physiognomy, or, The Corresponding Analogy Between the Conformation for the Features and the Ruling Passions of the Mind*, trans. Anonymous (London: np, 1827), 115.
50. Lavater, *Physiognomy*, 363.
51. Charles Augustin Vandermonde, *Essai sur la manière de perfectionner l'espece humaine* (Paris: np, 1756), 111.
52. On the role of French medical practitioners in discussions of public health, sex, and heredity, see Sean M. Quinlan, *The Great Nation in Decline: Sex, Modernity, and Health Crises in Revolutionary France c. 1750–1850* (Aldershot, Burlington: Ashgate, 2007); see also Anne C. Vila, *Enlightenment and Pathology: Sensibility in the Literature and Medicine of Eighteenth-Century France* (Baltimore: Johns Hopkins University Press, 1998).
53. For an overview on Frank and proto-eugenic theories, see Sara Eigen Figal, *Heredity, Race, and the Birth of the Modern*, 105–127.
54. On Frank and eighteenth-century ideas on racial improvement, see Sara Eigen, "Policing the Menschen-Racen," in *The German Invention of Race*, ed. Sara Eigen and Mark Larrimore (Albany, NY: State University of New York Press, 2006), 185–202.
55. Johann Peter Frank, *System einer vollständigen medicinischen Polizey* (Mannheim: Schwan, 1779), 457.
56. Felicity Nussbaum, *The Limits of the Human: Fictions of Anomaly, Race, and Gender in the Long Eighteenth Century* (Cambridge: Cambridge University Press, 2003), 25.
57. Nussbaum, *Limits*, 256.

8

ANALOGY OF ANALOGY

Animals and Slaves in Mary Wollstonecraft's
Defense of Women's Rights

PENELOPE DEUTSCHER

> The disgust at transgressed boundaries—animal/human; respectable/squalid; well-bred/low-bred—is irrepressible; yet, with Jemima, Wollstonecraft presses hard at these divisions, determined to test them further: "I was the filching cat, the ravenous dog, the dumb brute," Jemima says of herself, "I had not even the chance of being considered as a fellow-creature."
>
> —Barbara Taylor, *Mary Wollstonecraft and the Feminist Imagination*

The many works reviewed by Mary Wollstonecraft for the journal *Analytical Review* included Catharine Macaulay's *Letters on Education* (1790) and *The Interesting Narrative of the Life of Olaudah Equiano, or Gustavus Vassa, the African Written by Himself* (1789) but also Stanhope Smith's *Essay on the Causes of the Variety of Complexion and Figure in the Human Species* (1787), William Smellie's *Philosophy of Natural History* (1790), and an abridged version of Buffon (1792).[1] Several commentators have included these reviews among the many influences on Wollstonecraft's subsequent production of *Vindication of the Rights of Woman*, and her later work. Their arguments have opened the door to an evaluation of the extent of Wollstonecraft's response to those natural history arguments of the day with which, in the format of volumes for review, she briefly engaged.

For example, Daniel O'Neill, including this reference in the context of a broader argument concerning Wollstonecraft's debt to and engagement with the Scottish Enlightenment, has argued, as does Scott Juengel, that Wollstonecraft possibly responded to arguments stressing the importance of environmental factors on human psychology and physiology as analyzed by Stanhope Smith.[2] The role of environment on the form of the races and the sexes is also thematized by Catharine Macaulay, a writer much admired by Wollstonecraft[3] whose defense of equal education for women predated the latter's. In thinking about Wollstonecraft in the context of such influences, I ask also how her interest in the impact of environment on the development of physiological form and character interconnects with the role simultaneously played by what Wollstonecraft calls the "argument from analogy." Wollstonecraft specifically makes reference at one point to the "anatomists" of the day "who argue by analogy from the formation of the teeth, stomach, and intestines,"[4] and there is some emergent research concerning her interest in analogy as a means of argument.[5] Some attention has been directed to the role of analogy in theories of the development of living forms. Analogies between those deemed subordinated (women, slaves, animals, the "third estate" of pre-revolutionary France) were also common in the day. My argument is that the best reading of Wollstonecraft's use of analogy will interpret these influences in tandem. In this way we see better the flexibility of reference given to analogy in her work. Wollstonecraft moves across analogies between states of development or retardation, social conditions, properties, conditions of development, and claims to perfectibility, all working to accomplish complex parallels between the social and political claims of those she depicted as subordinated.

ENVIRONMENT AND ANALOGY: SAMUEL STANHOPE SMITH AND CATHERINE MACAULAY

In his *Essay on the Causes of the Variety of Complexion and Figure in the Human Species* Stanhope Smith argued that environmental influences produce femininity, whether it be the variations seen in the unpolished rural woman or in the urbane, graceful, cultivated woman. This point is used to mock the argument that different races or peoples should be seen as different species. Rebutting polygenism, as did Wollstonecraft, Smith stresses the difference:

> [B]etween the uncouth features, and unpliant limbs of an unpolished rustic, and the complacency of countenance, the graceful figure, and easy air and movement of persons in cultivated life! between the shaped and meaning face of a well bred lady, and the soft and plump simplicity of a country girl!

> We now easily account for these varieties which have become familiar to the eye, because we see the operation of their causes. But if we should find an entire nation distinguished by a composition of features resembling the one, and another by the contrary they would have as fair a title to be ranked under different species by certain philosophers as the German, and the Tartar.[6]

The passage is a satirical rejection of the distortions leading some to erroneously identify different human species. Smith claims that environmental factors can produce such differences among human peoples, as to stimulate this false interpretation. Moreover even within one people, the various environments in which women form (rural and aristocratic, for example) are so different as to produce a similarly distorting effect—one might almost have concluded, he ironizes, that cultivated and uncultivated, rich and poor women belonged to different species.

One other discussion of women in his work arises in a long piece on Native Americans in an appendix. These are references to women "destined by the customs of savage life," to "labors of drudgery,"[7] labors worsened by the "indolence" of savage men. The savage state is described by Stanhope Smith as not producing the improving qualities of compassion and sympathy:

> The savages of America, still more rude, and conversing only with the wildest scenes of nature, know nothing of those finer feelings of the heart, and that soft interchange of affections which give birth to the sentiments of compassion and sympathy.[8]

These savages are said to take pleasure in subjecting "inferior animals" to cruel torments.[9] It is unclear how the status of sexual difference is understood in this state. Smith's view is that a graceful and sweet, "European" femininity is the product of environmental factors, thus it is not surprising that he views women in the savage state as not greatly different to men in their tendency toward brutality. Both sexes are described as capable of torture, rage, anger, and violent retribution.[10] Neither the savage woman nor the man in these circumstances is attributed with the compassion he associated with the civilized human. So, a possible influence on Wollstonecraft proposed by O'Neill and others, proves to be one where the environmental factors highlighted by Stanhope Smith are attached to a representation of the savage, as entirely lacking in compassion: the wildness of nature in which he or she is immersed seems not to stimulate it.

If we turn to Catharine Macaulay's *Letters on Education* (1790), more widely recognized as an influence on Wollstonecraft's work,[11] empathy and compassion are again not identified by her as shown by humans in a savage state. For Macaulay, the cultivation of empathy and compassion is critical to the development of what is valuable in humanity. But in making the point, Macaulay seems to analogize the worst of the human with a state to which animals also can be driven. She writes of how beasts of prey will devour each other "when their appetites are in motion"[12] for their gratification, and connects this to the fact that "man, in the early ages of society, fed on man; and there is no violence which this being, who boasts that he is governed by reason, has not committed against his own species."[13] Men, when properly educated and raised so that their benevolence is cultivated, can and should be the "gentlest of animals."[14] But Macaulay is concerned that contemporary humans are instead tending toward inhumanity.

The instances of inhumanity evoked often concern the negligent or vicious treatment of animals, creatures of all kinds, ranging from pets, domestic animals, farm and wild animals, insects, reptiles, and even beasts of prey.[15] As these accounts are presented, the rhetorical function of references to the violent, early stages of human peoples is unclear. Macaulay does depict "savages" as brutish, without seeming to want to see brutishness as original and natural.[16] Savages may also be considered to have the potential for an original sympathy from which they, too (just as Macaulay says of contemporary humans), would have been turned by custom and habit. In her discussion, the comments move flexibly between early and contemporary states of humanity. Contemporary humans have been rerouted from their natural or potential gentleness—for this needs cultivation, practice, education, example, habit, and repetition. But when we are told of early stages of society, in which humans are said to have fed on humans, and to have committed every violence against their own species, this is followed by a discussion that slowly dissolves (as is clear in the reference to the treatment of African slaves) to a discussion of contemporary women. Thus the above reference to the primitive is followed by the remark, "What atrocious cruelties has not pride, the lust of power, riches, beauty, and the dire passion of revenge, given birth to!"[17]

Then Macaulay switches to an account of how it may instead be the arrogant recognition of superiority that motivates human violence toward the weaker (the example here is a hare): "And even where these keen excitements have been wanting, the mere insolence of superiority, and the force of habit, have given birth to injuries similar to those now suffering [sic] by this hare."[18]

This reference dissolves into a reference to slavery; thus, Macaulay moves from the treatment of the hare to that of the African slave: "Not to mention the treatment given by some of own (*sic*) countrymen to their African slaves!"[19]

Then she moves the reference back immediately to a description of ancient peoples, who appear to have the potential for sympathy but nonetheless indulged in cruelty to their fellow humans: "The Spartans, a race of men not destitute of the qualities of the heart, actually hunted the Helots in their sports."[20]

With this reference to the former practices of humans, Macaulay provisionally represents humans as originally cruel and barbarous peoples, not unlike beasts of prey. In one more reference to the contemporary human tendency to cruelty, violence, and inhumanity, she declares of humans: "they will never amount to positive excellence till all our barbarous customs are abolished."[21]

However, it is unclear that the primitive is being represented as the emblem of the cruelty the civilized human must abolish in the progress of perfectibility. It seems just as much that the "civilized" and "primitive" man are being seen as similarly cruel, while both possess the potential for positive excellence, assuming the cultivation of benevolence. There is an insinuated connection between the savage and an absence of compassion. But when Macaulay challenges the falseness of the man "who boasts that he is governed by reason,"[22] yet commits extremes of violence, she is referring as much to contemporary man as to the savage. It is contemporary man who is said to cruelly hunt for sport,[23] to be driven to atrocious cruelties either from insolence,[24] or pride, revenge, and the lust for power, riches, or beauty.[25] It is in this sense that he will be deemed like the savage.

Macaulay decries a parallel absence of compassion in the treatment of the "negro race," referring to the cruelty with which those in the West Indies were being treated.[26] At one point, she associates women with slavery, as when she acknowledges that "some degree of inferiority, in point of corporal strength, seems always to have existed between the two sexes,"[27] and claims that "this advantage, in the barbarous ages of mankind, was abused to such a degree, as to destroy all the natural rights of the female species, and reduce them to a state of abject slavery."[28] Again, for contemporary men to treat women similarly will be to treat them as do brutes and savages.

But while women's condition is analogized to that of slavery, it is not analogized to the condition of frail animals at the mercy of men's tendency to mistreatment. In fact, given the description of the claim made by animals, Macaulay might have concluded that women, physiologically weaker than men, had a claim to make, like animals, on men's benevolence. Although she does reject the view that the difference in strength merits the inequality in their education, she does not relate this to an additional analogy with animals. Here, there are two analogies: in the one association animals and slaves are both presented by Macaulay as making a claim on men's benevolence. While this particular analogy does not, in Macaulay's hands, extend to women, in the other association women are said to have been reduced to a state akin to that of slaves.

OLAUDAH EQUIANO: SLAVERY AND ANIMAL ANALOGIES

Like Macaulay's *Letters on Education*, Wollstonecraft's *Vindication of the Rights of Woman* (1792) was published at the time of the British parliamentary debates about slavery. Moira Ferguson[29] and Barbara Taylor[30] are among those who have discussed the resonance this gives to the repeating analogies suggested by Wollstonecraft throughout *Vindication of Rights of Woman* between women's legal and social condition and the state of slavery.[31] Attention is also drawn to a further work Wollstonecraft reviewed for *Analytical Review*, among the works in which she certainly had a special interest, Olaudah Equiano's *The Interesting Narrative of Olaudah Equiano or Gustavus Vassa* (1789).[32] Here again, as was common in slave narratives and antislavery rhetoric, analogies are deployed—in this case between slaves and animals. Using analogy to decry oppressive degradation, Equiano consistently analogizes the slave with the animal, as when the slaves are "all pent up together like so many sheep in a fold":[33] "cut and mangle[d] . . . and altogether treat[ed] in every respect like brutes";[34] "working in the fields like beasts of burden";[35] and "humbled" and again, "degraded" to "the condition of brutes."[36] He jolts the reader by stressing that not only are slaves treated as subhumans, but owners also claim or are attributed an aberrant status too extremely "above" the human:

> Surely this traffic cannot be good, which spreads like a pestilence, and taints what it touches! which violates that first natural right of mankind, equality and independency, and gives one man a dominion over his fellows which God could never intend! For it raises the owner to a state as far above man as it depresses the slave below it; and, with all the presumption of human pride, sets a distinction between them, immeasurable in extent, and endless in duration![37]

Here, Equiano's analogy functions almost as a literal equivalence. His narrative describes the literal, rather than analogical, purchase, branding, working, and herding of humans. Consider, by contrast, the antislavery sympathies expressed by Wollstonecraft in her review of Equiano. She wrote in a context when antislavery arguments, depicting slavery as inhuman, unchristian, brutal, inefficient, and expensive, competed with arguments defending slavery with reference to the subhumanity of some peoples.[38] Literary production, or the demonstration of literacy, mathematical, navigational, or accountancy abilities by slaves and former slaves could be, as Wollstonecraft noted of Equiano, powerful demonstrations that their humanity was not inferior:

> It has been a favorite philosophical whim to degrade the numerous nations, on whom the sunbeams more directly dart, below the common level of humanity . . . the activity and ingenuity, which conspicuously appear in the character of Gustavus [Equiano], place him on a par with the general mass of men.[39]

A common (and, some abolitionists argued, anti-Christian) view that certain peoples were subhuman was refuted with the argument that the "general mass" of Europeans ought to be considered the benchmark with which slaves were "on a par."

These comments occur in the context of widely circulating political analogies where references to animal-like treatment are commonly used to condemn the treatment of slaves. One can also turn to the use of analogies among French revolutionary thinkers suggesting that peasants had been treated like slaves, and, like animals. Writers who claimed women's rights—among some examples, Astell in the previous century, Macaulay and Condorcet[40]—also leaned toward the parallel of women's legal and social conditions and those of slaves.

Insofar as the use of analogy in these contexts was common, I direct closer attention to the form that this kind of analogical thinking takes in Wollstonecraft's work. Reiterating widespread associations with apparently powerful rhetorical appeal, Equiano and Macaulay bring together the slave and the animal; Stanhope Smith and Macaulay the savage and the brute; Macaulay and a number of others the slave and the woman. But in Wollstonecraft's writing, *slave, brute, animal* (and in fact, we can add *child, savage, plant*[41]), and *woman* are all analogized together, adding layers of complication.

Analogy certainly plays a role, for example, when Taylor notes the (for Wollstonecraft) alarming blurring of boundaries occurring when

Wollstonecraft presents as abhorrent the proximity of the European husband to the vicious West Indian slave owner, who is in turn analogized with the savage. Taylor's way of putting this point is that this appeal to disgust "carried new ideological baggage," insofar as it enfolded a number of layers of association. When the husband is analogized with the slave owner, this is rhetorically effective because this author assumes the slave owner is already considered a repellent figure.[42]

How might we further understand the terms and language of this multilayered appeal to existing antislavery sentiment? When the association (husband = slave owner) capitalizes on the capacity to appeal to the contemporary reader's presumed disgust at the slave owner, what is also occurring is a redoubled analogy. One sees, in other words, a deployment of analogy of analogy. Wollstonecraft will tell us that the husband is disgusting if "like" the slave owner, because the slave owner is already disgusting because "like" a "savage" or "barbarian." If her readers were disgusted by slave ownership, so, too—this analogy implies—they should be repulsed by the legal privileges of the European husband. Thus, we can add to the disgust at what ought not to be in proximity the role both of analogy, and of analogy of analogy, in that sequence.[43]

But Wollstonecraft's use of analogy requires considerable parsing, as seen in her review of Equiano's slave narrative. Here, Wollstonecraft certainly decries those who degrade certain peoples as inferior. But, as was not uncommon in abolitionist rhetoric, she does not deny that some peoples have been made by nature "inferior to the rest of the human race." Her claim is instead that inferiority is no justification for slavery. If anything, then, Wollstonecraft takes the trouble to stress that Equiano's literary skills do *not* manifest "extraordinary intellectual powers." Instead, she says, they may be analogized to the powers of those who "fill the subordinate stations in a more civilized society." More slyly, they are analogized by negation to the European mechanic.[44] In other words, Wollstonecraft's expression of antislavery sentiment was consistent with her also supposing the subordination and inferiority of certain classes of humans, but this view, too, was secured by analogy.

WOLLSTONECRAFT AND THE ANIMAL ANALOGY

My argument is that while the analogy between women's subordination and slavery, if not uncommon in feminist thought of the eighteenth and nineteenth century, is hardly uncomplicated, Wollstonecraft's analogical play of associations, integrating questions of human-animal proximity, human-plant[45] proximity, and of the relations between human peoples in different states and roles (children, savages, servants) renders the account of likeness with the slave more complex still. As Wollstonecraft speculates about the relations between different human peoples, states, habits, and a number of different animals, and plant forms, one can, as commentators have separately suggested, think of the influence of figures from Macaulay to Stanhope Smith, Equiano, and the analogies of Linnaeus and Buffon.[46] But it is Wollstonecraft who, drawing on a range of arguments concerning the formation of animals, races, peoples, and plant life, arguments opposing slavery, and arguments concerning sexual difference, also draws from these resources a redoubling form of analogical argument not seen with the same degree of compression in Stanhope Smith, Macaulay and other figures considered as possible influences.

Just a focus on the animal references alone in *Vindication of the Rights of Woman* offers an excellent example of the analogical flexibility in Wollstonecraft's work. There are, first, a number of physiology-oriented analogical references to animals in her work, which arise in the context of her objections to indolence, seen as both physiologically and morally harmful:[47]

> Throughout the whole animal kingdom every young creature requires almost continual exercise, and the infancy of children, conformable to this intimation, should be passed in harmless gambols, that exercise the feet and hands.[48]

As seen here, animals could serve as a point of reference for human physiological necessity: as such parallels between humans and animals enhance—rather than undermine—her account of human developmental requirements. Via this argument, Wollstonecraft argues that women must be vigorous and active, like every creature in the animal kingdom, and indeed, like plants: we're also told that women should not "languish like exotics"[49] constrained by artificial environments, their modesty becoming a "sickly hot-house plant,"[50] these stifling environments promoting in women an unnatural weakness in both body and mind.

So Wollstonecraft could positively stress the animal analogy (accompanied by some associated plant analogies) in order to make some claims on behalf of women. But, the role of analogy functioned very differently—here it would be negative, not neutral—when Wollstonecraft made claims on behalf of women so as to elevate them over the animal. Women would be definitively elevated over the animal by the exercise and development of their rational capacities and moral sense. The inhibition of these faculties reduced them to the domains of sensibility and sensuality. Being endowed with a soul yet not fully guided by reason, rendered women closer to the animal. She referred to the contemporary debate about whether "the dignity of the female soul [was] disputable as that of animals."[51] Second, then, Wollstonecraft directly argued that the occupations women could access were so trivial as to render them hardly better than docile, domesticated animals.[52] Women were "advised to render [them]selves gentle, domestic brutes."[53] She associates the possibility of women having been made for man with an imagery of them "patiently bit[ing] at the bridle."[54] When their occupations and passivity are described as animal-like, they are likened by Wollstonecraft to ornamental and pet animals: birds in cages and lapdogs. "Is it surprising," she asks, that some of them "hug their chains, and fawn like the spaniel?"[55] Another sense of the animal-like is associated with the woman who is, for her husband, the patient drudge, who fulfills her tasks as does a blind horse in a mill.[56] Wollstonecraft does consider that "brutes" were "principally created for the use of man,"[57] and so to treat women in such a fashion is to render them, in this sense, like brutes. Here, the analogy functions to set a low mark of what it is degrading for women to be "like."

Third, Wollstonecraft also profits from the fact that the term *brute* can be used to refer to the animal realm (the brutes), to uncultivated humans (brutish), and also, as we shall see, to callous and cruel humans (again, brutish). This gives two more senses in which women can be said to be brutish (where to be brutish is not the same as being like a brute, nor, evidently, is it the same as being treated like a brute). Wollstonecraft depicts the moral obtuseness of women who may have a narrowly focused affection for their children at the expense of a broader and more balanced perspective oriented to all those who have a claim on their interest, compassion, and humanity. Their limited outlook arises from their deprivation of an education and culture that promotes the principle of justice and collective social interest:

> The affection of some women for their children is, as I have before termed it, frequently very brutish; for it eradicates every spark of humanity. Justice, truth, every thing is sacrificed . . . and for the sake of their own children they violate the most sacred duties.[58]

Fourth, a parallel to Macaulay's focus on benevolence is present in Wollstonecraft's work. Wollstonecraft agrees about the far-reaching importance of humane treatment of animals: "Humanity to animals should be particularly inculcated as a part of national education, for it is not at present one of our national Virtues."⁵⁹ And she even agrees that "justice, or even benevolence, will not be a powerful spring of action, unless it extends to the whole creation."⁶⁰

But the use of the reference to animality is more complicated—used analogically—than it is in Macaulay's *Letters on Education*. Wollstonecraft explains the subordination of women in terms of a history according to which "brutal force has hitherto governed the world."⁶¹ Here, men partake in, or revert to, the realm of the brutes in rendering women poor brutes. Treating women as lapdogs or birds in cages might appear to belong to the elegant *moeurs* of the cultivated European man. Yet Wollstonecraft argued that this behavior also showed a tendency in men toward the brutish. At this point Wollstonecraft amplified a second level of analogy: women were treated like animals and also like slaves, and insofar as they could be analogized to slaves,⁶² they were likened to those whose treatment was also analogized to that of animals.⁶³

Wollstonecraft uses references both to being on a chain (the woman said to be like the spaniel), and in chains (the woman presented as like the subjugated slave). Thus, likening European men to slave owners she laments: "Would men but generously snap our chains."⁶⁴ Playing again across two senses of "slavish," she claims that girls are both "taught slavishly to submit to their parents," and that they are thereby "prepared for the slavery of marriage."⁶⁵ And most literally: "Is one half of the human species, like the poor African slaves, to be subject to prejudices that brutalize them, when principles would be a surer guard only to sweeten the cup of man?"⁶⁶

With respect to the slave analogies, women are also presented extensively as slaves in another sense. Enslaved by their upbringing, they are slaves to their bodies. Lacking the training necessary for independence of thought, they are also slaves to sensuality. Thus she speaks of those who "enslave women by cramping their understandings and sharpening their senses."⁶⁷ As a result women are "slaves to their persons, and must render them alluring."⁶⁸ They are "in the same proportion, the slaves of pleasure as they are the slaves of man."⁶⁹ Again, redoubled senses of slavery interconnect—it is an indication of women's enslavement to men and the social expectations of femininity, that they are encouraged to lead child- or animal-like lives dominated by the sensuous and an absence of rational principles. Referring to marriage as the only way women can rise in the world (or, in many cases, survive), she writes that the very desire to marry makes mere animals of women, and she goes on to interconnect this animal-likeness with childishness: "when they marry, they act as such children may be expected to act: they dress; they paint, and nickname God's creatures."[70] Wollstonecraft moves from the animal analogy for slavery, via a child analogy, to an analogy with the putatively less progressive *moeurs* of some foreign peoples when she describes women as "educated in worse than Egyptian bondage."[71] So, when she denounces the hierarchy between women and men, she declares (and here we can itemize the analogies in play: birds, lapdogs, insects, children, Egypt, Orient): "Surely these weak beings are only fit for the seraglio!"[72]

CLUSTERING ANALOGIES: ANIMALS, SLAVES, WOMEN

At the point at which Wollstonecraft is using analogies of analogies, as in the series woman-slave-animal, a countering force to her own analogies begins to be apparent in her work. Although Wollstonecraft certainly uses the image of the savage—and the negative image of being closer to the brute—to decry the European man's treatment of women,[73] she also is willing to suggest that the savage can be above, rather than below, the European. Like animals and children, the lower classes are also enmeshed in this analogization with the savage:

> Tenderness for their humble dumb domestics, amongst the lower class, is oftener to be found in a savage than a civilized state. For civilization prevents that intercourse which creates affection in the rude hut, or mud cabin, and leads uncultivated minds who are only depraved by the refinements which prevail in the society, where they are trodden under foot by the rich, to domineer over them to revenge the insults that they are obliged to bear from their superiors.[74]

The savage has been representing functions, states, and qualities considered below a European level of development. But here, and despite her quarrels with Rousseau, Wollstonecraft proposes that the supposed refinements of European civilization corrupt what might otherwise be a simple, natural, affective primitive state: one in which the proximity of the "hut" or "cabin" might induce compassion or warm affections. This is not the only case in which Wollstonecraft suggests that the low might be high. Like Macaulay she also presents some animals as offering (in specific senses) preferred models of behavior. Some women have become so denatured that they do not even reach the instinctive "decency" shown by many "brutes" (animals).[75] So, on some points, animals could be positioned above humans, offering to humans a higher standard. On the other hand, humans can be *reduced* to the subhuman level of acting like brutes. And indeed, they can be brutes in their treatment of brutes.[76]

Wollstonecraft also claims that being trodden underfoot (i.e., being treated like a brute) inclines humans to become more like brutes. Here, the analogical cluster again unravels and reforms. Now, cruelty is brutish, and animal brutes fall victim to brutish humans. But those violent humans, treating animals cruelly, have become brutish themselves by being treated like brutes: "This habitual cruelty is first caught at school, where it is one of the rare sports of the boys to torment the miserable brutes that fall in their way."[77]

Here Wollstonecraft suggests an extension from the boy's treatment of the brute, to the man become accordingly both brute and barbarous, prone to brutish treatment of the brute, and accordingly brutish also in the treatment of women (who, as we know, will thereby themselves become poor brutes). As she says, more simply, of boys treated by other boys *as* brutes, and accordingly treating animals with cruelty: "The transition, as they grow up, from barbarity to brutes to domestic tyranny over wives, children, and servants, is very easy."[78]

Wollstonecraft thus extends an argument to which Macaulay also appeals, concerning the chain reactions of cruelty and a "becoming-brute," thus including the phenomena of subordination and the outward expression of violence and so tyranny against women.

But although women are analogized in the midst of this discussion to poor harmless brutes, Wollstonecraft also considers that women can be rendered brutes in a different sense. Consistent with the arguments, which appear to imply a passage of behavior (if one is likely to treat other creatures as one has been treated, brutish treatment will create the brute who treats others brutally), Wollstonecraft expresses concern about the quality of treatment by women toward the creatures around them. In this respect there are two problems. Women can be cruel and indifferent to weaker creatures, and, lacking the inculcation of rational or moral principles, they can also be partial and arbitrary in their treatment of them.

As a consequence, we do hear of the woman who extends her concern to some forlorn animals (a suffering bird, ox, or ass). But it will be recalled that according to Wollstonecraft "justice or even benevolence" must be "extended to the whole creation." The problems arise when the woman fails to extend her concern for small suffering animals to concern for other animals for which she is responsible (e.g., her horses) or, indeed, neglects her servants. This scenario is said to be all too common:

> The lady who sheds tears for the bird starved in a snare, and execrates the devils in the shape of men, who goad to madness the poor ox, or whip the patient ass, tottering under a burden above its strength, will, nevertheless, keep her coachman and horses whole hours waiting for her, when the sharp frost bites, or the rain beats against the well-closed windows.[79]

Or her brutishness may be seen as she elevates her concern for her pet dogs over her concern for her children. Women may also be described as brutish when they are selfishly and, in an insular sense, concerned only with their children at the expense of all else (particularly the claims of justice). But they are also said to be brutish because they are selfishly and inwardly concerned with their animals to the exclusion of their children.

Having intermittently analogized animals with slaves, and mistreated, subjected women with the women of the harem, Wollstonecraft remarks:

> This [mistreatment, this absence of benevolence] . . . shows how mistaken they are who, if they allow women to leave their harems, do not cultivate their understanding, in order to plant virtues in their hearts.[80]

Women are subordinated slaves, and slaves are now said not only to be mistreated but—without cultivation and education—to lack morality and benevolence. So, they are (analogically) enslaved and accordingly they are prone to mistreat creatures under their power. For Wollstonecraft these ideas go together. They are treated unjustly, and they approach the world with an absence of the principles of justice or little capacity to follow their dictates—so they act unjustly.

This suggests at least two reasons why what should be "lower" in some respects can be higher. It can be a matter of valuing the instincts from which we should not become denatured. If French women have been badly corrupted by culture, the brute may be said to have, at least comparatively, more instinctive prudishness. We are also told that both the lower classes and the "savage" may show more tenderness toward their humble, dumb domestic animals than do civilized or more refined, *mondaine* women and men who do not benefit from the more intimate social intercourse of huts and cabins.

We have seen other reasons why women are considered to lack what would seem to be natural affection. It must be learned, cultivated, made a matter of principled habit. Without it (and this is consistent with earlier remarks that animals may also be favored over children) women may dangerously neglect their infants.[81] Again, this renders women, in this respect, beneath the brutes. Here, again, it seems the natural instincts (of affection and care) are lost through the effects of a corrupting culture, as when women become excessively responsive to flattery. As a consequence, the brute can be "higher" than the woman. But even though "savages" are described as closer to their instincts and to innocence in a positive sense, and so closer to the animal, savages are also associated with a childishness and superficiality to which men and women ought not aspire. Here the savage is the benchmark, not of natural and innocent inclinations valued by Wollstonecraft, but of immoderation and unregulated passion. Now they are the marker of the low level to which women are similarly reduced (women = savage), if they do not cultivate morality, rational principles of justice, and benevolence:

> An immoderate fondness for dress, for pleasure and for sway, are the passions of savages; the passions that occupy those uncivilized beings who have not yet extended the dominion of the mind, or even learned to think with the energy necessary to concatenate that abstract train of thought which produces principles. And that women, from their education and the present state of civilized life, are in the same condition, cannot, I think, be controverted.[82]

So, both the animal and the savage work doubly in Wollstonecraft's *Vindication*, both serving as the marker *to which* uncultivated humans can sink (as when women act so brutishly as to be likened to "savages"), and the marker *below which* humans can sink (women act so brutishly, that savages or animals are above them). We are told that the European ought to be the benchmark of the animal or savage, but that the animal or savage may be better than the human in animal "decency" (or in the *savage*'s intimate affection in mud huts). The savage will be both importantly better than—and the low benchmark of—the thwarted perfectibility seen in women's uncultivated humanity.

This gives an extreme rhetorical flexibility to what will count as brutish, and what will count as "savage": both the brute savagery of men's violence toward women and the perception of European women as lamentably in proximity with the savage insofar as they are uncultivated or vain. As slave, savage, and class come to overlap, we see Wollstonecraft evoking the superficial values associated with the primitive simplicity of Africans subjected to slavery:

> So far is the first inclination carried, that even the hellish yoke of slavery cannot stifle the savage desire of admiration which the black heroes inherit from both their parents, for all the hardly-earned savings of a slave are commonly expended in a little tawdry finery.[83]

Race and social rank are then allowed to overlap in an analogically presented account of the preferences of the European servant and the African slave (superficiality and trivial satisfaction). Thus the sentence goes on: "And I have seldom known a good male or female servant that was not particularly fond of dress. Their clothes were their riches."[84] Only to lead to the association between African slaves, servants, and women considered as a group: "*and I argue from analogy*, that the fondness for dress, so extravagant in females, arises from the same cause—want of cultivation of mind."[85]

In sum, it is true that many of the analogies and associations to which Wollstonecraft appealed were common coin, certainly in many of the works she reviewed—most obviously, the association between women's status and slavery promoted by many women's rights defenders.[86] Stanhope Smith had seen savage women as subjected to drudgery by savage men. Macaulay had claimed that contemporary men continued to favor some barbaric habits in modern times, particularly seen in their treatment of animals and slaves.[87] It is in Wollstonecraft's work, however, that all these associations cluster together.[88]

Thus, to follow one of the ways in which this cluster takes shape, cruelty is savage, savages treat women badly, men remain in both these senses barbaric, women are like animals in being so treated, subjugated women are both like slaves and like animals, and men are not only like savages but also like brutish animals in treating women like animals. Wollstonecraft, in other words, clustering her analogies, used analogies that were themselves analogies of analogies. I want finally to think about the role played by this clustering.

ANALOGIES AS ANALOGIES OF ANALOGIES

When Wollstonecraft drew on the rhetorical possibility of analogizing the condition of women with that of slaves, the abolitionist movement already appealed to analogy—particularly the analogy between the treatment of slaves and animals. In the case of Wollstonecraft, I have considered the additional effect of what I have described as "analogy of analogy."

Equiano could point out a number of respects in which slaves were literally treated as animals. Perhaps a conversion of this analogical context occurred when it repeated in the affirmation of women's rights. Arguably, the defender of women's rights, who likened women's conditions to those of the animal or the slave, indicated the nonliterality of the analogy. The reader is invited to think that the (middle class) woman (if she is the default woman's rights claimant[89]) evidently should not be like enslaved peoples, nor like animals—a view that blurs with the insinuation that they should not be treated like (putatively) inferior peoples and creatures. It is through that insinuation that the writer undertakes to startle with the analogy. As a result, and perhaps unlike the tone and effect of Equiano's analogies, perhaps when some women's rights claimants analogized themselves (women = slave) to what was already in circulation as an analogy (slave = animal), the analogy drew attention to itself. It winked—asserting that the literal version would be improper and was not true.

Some may wish to argue that the legal status of British women in the eighteenth century with respect to the legal rights of their husbands over them, their children, property or earnings, was so bad as to amount to literal slavery. If one should nonetheless resist this view, it is not just for reasons of an interpretative politics committed to stressing specificity; though arguably, the historical differences between the legal status of women, between women of different classes, and slaves, matter as much as the likenesses. But the claim that "women" were being treated "like slaves" *also* de-emphasized the association of middle class European women with women of different cultures, peoples, and stations in life. European women were both likened to and simultaneously held apart (because of the insinuation of aberration) from the association with enslaved women, women in "seraglios," and the "servant class."[90] These were providing categories of abjection to which European women were to be understood as *reduced*. The European middle class woman's claims derive a particular rhetorical profit from the suggestion that they were being treated "like slaves." This surplus is lost in an account that is willing to reliteralize the analogy. That women and slaves were not engaged in the same legal or rhetorical battle is clear from the absence of equivalent arguments that slaves were being treated "like women." The reference to the slave, with the enfolded association with animality, grounded the claim that women were being degraded by being treated "like slaves." This is the difference between Equiano's "like brutes" and Wollstonecraft's "like brutes."

In sum, tracking the use by Wollstonecraft of analogies for which a number of contexts could be identified, commentators have identified among the likely influences on Wollstonecraft's work figures including Stanhope Smith, Buffon, Linnaeus, Macaulay, and Equiano. Thus it could be said that the analogical rhetorical gesture was contextually available to her—widely used in pro- and antislavery defenses, in debates about natural rights, and in natural history. Yet in Wollstonecraft's hands, there is a use of an analogical form that seems to be more specific. It is a conversion resulting from the concurrent, interweaving rhetorical parallels between slaves and brutes, women and children, women and savages, women and servants, women and plants,[91] women and slaves, a constant redoubling that coincided with the insinuation by analogy of aberration: Wollstonecraft's analogies claim that women should not be like birds, oxen, horses, hothouse plants, insects, slaves. She does draw attention to her use of analogy, in the sense that we will find several references to the term, and the self-reflexive statement, echoed several times in Wollstonecraft's *Vindication*: "and I argue from analogy." But if, in this way she could be said to draw attention, albeit quietly, to this means of reasoning, she appears not to think it necessary to make a direct case for its usage. Above all, she does not draw attention to the extent to which she is, in fact, using analogy of analogy, or to put it another way, analogical conversion: from analogy to the analogy of the analogy as seen most simply in the compressed sequence: animal-slave (like-animal)-woman (like-slave-like-animal) and so on.

Wollstonecraft mentions the argument from analogy, and she also looks right through it. She does not suppose that her argumentative *form* (in addition to her claims on behalf of women) requires defense. A high degree of self-reflexivity with respect to one's own rhetoric is—to be sure—rare in any context. But by placing it under scrutiny, one can consider more directly the rhetorical profit that is derived once she has a number of analogical figures in play: animal, savage, brute, slave, slave-owner, husband, woman. What we see is that this is not a series of equivalents. For when the claim that women are like animals and slaves (not to mention children and savages), serves the interests of women's claim to a better status, what links the analogy with the analogy of the analogy is the hinge of what may be named an indirect, aspirational, analogical subordination of those whom it would (according to these embedded subordinations) be degrading for women to be "like." To identify this component is not to question that Wollstonecraft considered slavery abhorrent. But it is to recognize that when she associated the claims of slaves and women, insofar as both were also likened to animals, savages, and children, she may have *both* reiterated a contemporary abhorrence of slavery but also reiterated a number of embedded subordinations.

NOTES

1. On the attribution of these reviews to Wollstonecraft see Janet Todd, "Prefatory Note," in *The Works of Mary Wollstonecraft*, ed. Janet Todd and Marilyn Butler (New York: New York University Press, 1989), vol. 7:14–18, 18; and Daniel I. O'Neill, *The Burke-Wollstonecraft Debate: Savagery, Civilization and Democracy* (University Park: Pennsylvania State University Press, 2007), who also provides references for the development of what he describes as a scholarly consensus on this point.
2. O'Neill, *The Burke-Wollstonecraft Debate*, 119. For a different discussion of the same influence (one that focuses on the influence of Stanhope Smith on Wollstonecraft primarily in her *Letters Written During a Short Residence in Sweden, Norway, and Denmark* see Scott Juengel, "Countenancing History: Mary Wollstonecraft, Samuel Stanhope Smith, and Enlightenment Racial Science," in *English Literary History* (*ELH*) 68 (2001):897–927. Wollstonecraft's engagement with natural history writers is not a widespread theme in commentary, but it has also been taken up by Sharon Ruston in "Natural Rights and Natural History in Anna Barbauld and Mary Wollstonecraft," in *Literature and Science*, ed. Sharon Ruston (Cambridge: D. S. Brewer, 2008), 53–71; in Anka Ryall, "A Vindication of Struggling Nature: Mary Wollstonecraft's Scandinavia" in *Mary Wollstonecraft's Journey to Scandinavia* (Stockholm Studies in English XCIX), ed. Anka Ryall and Catherine Sandbach-Dahlström (Stockholm: Almqvist & Wiksell International), 117–139; and in Alan Bewell, "'Jacobin Plants': Botany as Social Theory in the 1790s," *Wordsworth Circle* 20, 3 (1989):132–139. Ruston makes a case for the influence of Smellie and Buffon, but also mentions Wollstonecraft's reviews of Thomas Bewick and Ralph Beilby's *A General History of Quadrupeds* (1790), and John Rotheram's *The Sexes of the Plants Vindicated* (1790). Anka Ryall discusses the likely influence of Buffon and Linnaeus on Wollstonecraft's *Letters Written During a Short Residence in Sweden, Norway, and Denmark* (analyzed by Juengel as responding to Stanhope Smith, and also to contemporary theories of physiognomy) and proposes a reading of Wollstonecraft as responding to Linnaean botany.

3. See Mary Wollstonecraft, *A Vindication of the Rights of Men and a Vindication of the Rights of Woman and Hints*, Cambridge Texts in the History of Political Thought, ed. Sylvana Tomaselli (Cambridge: Cambridge University Press, 1995), 188.
4. Wollstonecraft, *A Vindication*, 82, n2. The passage begins with her rebuttal of Rousseau: "if benignity itself thought fit to call into existence a creature above the brutes, who can think and improve himself, why should that . . . gift . . . be . . . a curse?" She then comments "contrary to the opinion of anatomists, who argue by analogy from the formation of the teeth, stomach, and intestines, Rousseau will not allow a man to be a carnivorous animal. And . . . he disputes whether man be a gregarious animal" (Mary Wollstonecraft, *A Vindication*, 82, n2.). A further use of the term analogy is seen in a comment made against Lord Chesterfield's promotion, in his *Letters to His Son on the Art of Becoming a Man Of the World and a Gentleman* of an "early knowledge of the world" (188), a premature exposure that would in fact, she rejoins, poison a youth's expanding powers as they are taking form. Here she promotes a parallel with a fruit-tree's life cycle. It cannot give fruit while still in its sapling stage of development, and the vain attempt to bring forth its fruit early "exhausts its strength" such that it will not be able to "assume its natural form" (189): "Our trees are now allowed to spread with wild luxuriance, nor do we expect by force to combine the majestic marks of time with youthful graces; but wait patiently till they have struck deep their root, and braved many a storm. Is the mind then, which, in proportion to its dignity advances more slowly towards perfection, to be treated with less respect? To argue from analogy, every thing around us is in a progressive state" (191). (The progressive state to which she refers here relates to development within one generation.) This same discussion includes within its reference to the life cycle of a tree, a concurrent analogy to the cycle of the seasons ("who would look for the fruits of autumn during the genial months of spring," 189), and an analogy with metal cohesion, which embeds the concurrent analogies to plant *and* human formation seen above: "the vain attempt to bring forth the fruit of experience, before the sapling has thrown out its leaves, only exhausts its strength, and prevents its assuming a natural form; just as the form and strength of subsiding metals are injured when the attraction of cohesion is disturbed" (189). The abridged version of Buffon that Wollstonecraft reviewed for *Analytical Review* begins with the forces of attraction and repulsion, and Wollstonecraft's familiarity with Buffon raises the question of the extent to which scientific analogy, and

also the problematization of analogy in that context, might have been a consideration in her use of the term. The abridged version contains references to the technique of drawing inferences from the progress of diseases and remedies in horses to humans (89), an analogy in bulk between the hippopotamus, elephant, rhinoceros, and whale, and the consequent likelihood that they reproduce similarly (240); the likelihood that similarities of form in species (such as possessing a horn) will be matched with other similarities (for example, whether this attribute manifests in only one, or both sexes) (271). Thus the term is often used in the sense of predictive similarity (thus the "whole analogy of nature" is said to make it unlikely that toads could really live for two centuries in hibernation, 498), or simple questions of relationship and general likeness: thus, in a discussion of the cuckoo: "Brisson makes no less than twenty-eight sorts of them; but what analogy they bear to our English cuckoo I will not take upon me to determine," 383); in a discussion of crustaceans that their external armature is analogous to ossature (465). However, in other direct mentions Wollstonecraft makes in *Vindication* of her own use of analogy there is no particular indication that she is still thinking of the "anatomists." Wollstonecraft's third reference to her use of analogy is: "That woman is naturally weak, or degraded by a concurrence of circumstances, is, I think, clear. But this position I shall simply contrast with a conclusion, which I have frequently heard fall from sensible men in favor of an aristocracy: that the mass of mankind cannot be any thing, or the obsequious slaves, who patiently allow themselves to be driven forward, would feel their own consequence, and spurn their chains. Men, they further observe, submit everywhere to oppression, when they have only to lift up their heads to throw off the yoke; yet, instead of asserting their birthright, they quietly lick the dust, and say, let us eat and drink, for to-morrow we die. Women, I argue from analogy, are degraded by the same propensity to enjoy the present moment; and, at last, despise the freedom which they have not sufficient virtue to struggle to attain" (126). A further repetition of the declaration "I argue from analogy" is discussed later in this chapter; here an analogy is suggested between the love of "tawdriness" among "savages," servants, and slaves (285).
5. See the references to Bewell, Ryall, and Taylor, footnotes ii and xlii.
6. Samuel Stanhope Smith, *An Essay on the Causes of the Variety of Complexion and Figure in the Human Species* (New Brunswick: J. Simpson, 1810), 179.
7. Perhaps, given these references to the cultivated woman's graceful figure

and easy air and the rustic woman's rough but soft and plump simplicity, we can suppose Stanhope Smith distinguishes both from the more wretched conditions he associates with women in the "savage" state. However, we do not learn anything here of Smith's views about the status of contemporary European women.
8. Stanhope Smith, *Causes of the Variety*, 402.
9. Ibid., 401. As such, they are described as children: "There is, indeed, a kind of wantonness in cruelty which forms a part of the character of the American savage that resembles the pleasure which children are often seen to take in the writhings and convulsions of the inferior animals subjected to their persecutions and torments. A savage is, in many respects, little more than a grown child. But in the moment of victory and triumph, in their barbarous carousals, and the wild frolic of all their spirits and their passions, they are still more cruel and unreflecting than on other occasions, and derive a more horrible diversion from the miseries of their captives" (401).
10. With these various associations in play it is also notable that in discussing different peoples and races, he makes only a few references to animals and the question of environmental impact on animal species. He makes clear that the environmental argument does have a bearing on the forms and dispositions of animals, though with important differences given his view that animals cannot "work on themselves." At one point he does describe the "savage" as "a taciturn animal" (*Causes of the Variety*, 346). Because the argument he is making, as presented, would hold as much for the sexes, for different human peoples, and for animals, there is no particular rhetorical impact of stressing that humans, insofar as they are dependent for their form and qualities on environmental considerations, are "animal-like."
11. Ferguson observes that Wollstonecraft had not manifested special interest in arguments for women's status as slaves in *Vindication of the Rights of Man* (1790), prior to her reading of Macaulay. Moira Ferguson, "Mary Wollstonecraft and the Problematic of Slavery," in *Colonialism and Gender Relations from Mary Wollstonecraft to Jamaica Kincaid: East Caribbean Connections* (New York: Columbia University Press, 1993), 8–33, 17.
12. Macaulay, *Letters on Education*, 118.
13. Ibid., 119. Here there is a stronger analogy between the human in the early stages of society and beasts of prey but she qualifies that the violence is not entirely gratuitous, adding that it is seen, "whenever they [his own species] have been found in opposition to [man's] fancied interest" (119).

14. Ibid., 87. Arguing that "when properly educated, man is the gentlest of all animals" (87). Macaulay doesn't appeal to the option of representing violent humans as "like brutes."
15. Rather than likening humans in their violence to beasts, Macaulay proposes the analogy of wild gods to whose unpredictable powerful human whims subordinated animals are subject.
16. Macaulay considers violent humans to have been rerouted from their intrinsic unwillingness to hurt each other, and particularly hurt the weak.
17. Macaulay, *Letters on Education*, 119.
18. Macaulay, *Letters on Education*.
19. Ibid.
20. Ibid., 100.
21. Ibid., 119.
22. Macaulay, *Letters on Education*.
23. Ibid., 118.
24. Ibid., 119.
25. Macaulay, *Letters on Education*.
26. Ibid., 157.
27. Ibid. 129.
28. Macaulay, *Letters on Education*.
29. Ferguson, "Mary Wollstonecraft," 8–33.
30. Barbara Taylor, *Mary Wollstonecraft and the Feminist Imagination* (Cambridge: Cambridge University Press, 2003).
31. Taylor makes several contextualizing observations about these associations. First, there was already a tradition which dated back at least to the 1600s of depicting women's role as enslavement (Taylor, *Mary Wollstonecraft*, 226). The analogy had also been used by Mary Astell in 1700 (99). The Dissenter movement had used the imagery of slavery to denounce the encroachment on religious freedom. Moreover, Taylor mentions the association in some quarters of the European Enlightenment with a new understanding of women's role: "philosophical historians as diverse as Montesquieu, Helvetius, John Millar, and William Alexander traced women's rise from the brutal enslavements of barbarism, when they had been valued only as childbearers, through to the respectful affection accorded them in commercial societies" (156–157).
32. Equiano, also known as Gustavus Vassa, was a slave whose narrative describes having been kidnapped in 1756 at the age of eleven from Africa, sold by a planter to a naval officer, then to a merchant officer, from whom he eventually purchased his own freedom. He then worked

as a freeman on trading ships, settled in London and published an autobiography in 1789. See Vincent Carretta's introduction to the Penguin edition of Equiano's "Interesting Narrative," for further discussion of the status of the autobiographical facts recounted by Equiano/Vassa.
33. See Olaudah Equiano, *The Interesting Narrative and Other Writings*, ed. Vincent Carretta (Harmondsworth: Penguin, 2003), 60.
34. Ibid., 105.
35. Ibid., 109.
36. Ibid., 111 and 232. Equiano once offers the parallel to which Wollstonecraft also appeals: the slave owners are brutes in treating the slaves like brutes. Slave owners are, in this respect, no better than the "savages," "brutes," "barbarians," "Samaide" or "Hottentot" who might otherwise be considered lower: "That if any negro, or other slave, under punishment by his master, or his order, for running away, or any other crime or misdemeanor towards his said master, unfortunately shall suffer in life or member, no person whatsoever shall be liable to a fine; but if any man shall out of wantonness, or only of bloody-mindedness, or cruel intention, willfully kill a negro, or other slave, of his own, he shall pay into the public treasury fifteen pounds sterling. And it is the same in most, if not all, of the West India islands. Is not this one of the many acts of the islands which call loudly for redress? And do not the assembly which enacted it deserve the appellation of savages and brutes rather than of Christians and men? It is an act at once unmerciful, unjust, and unwise; which for cruelty would disgrace an assembly of those who are called barbarians; and for its injustice and insanity would shock the morality and common sense of a Samaide or a Hottentot" (109).
37. Ibid., 111.
38. Compare Equiano's use of the animal metaphor to that of the Danish naturalist Johann Christian Fabricius (as discussed in this volume by Renato Mazzolini in his chapter "Skin Color and the Origin of Physical Anthropology [1640–1850]"), who, in 1804, reaffirms the lack of intellect of blacks. As Mazzolini writes, Fabricius suggested they were to be understood both as a cross of white man and ape, they were then not the brother but the "half-brother" of the European, and, despite the fact that he was reacting to the slave revolts in St. Domingue, he affirmed that blacks were "flocks of sheep" "annually led" by whites "from the coasts of Guinea to the West Indies."
39. Mary Wollstonecraft, *A Vindication*, 100.

40. See Condorcet's "On the Admission of Women to the Rights of Citizenship" (1798).
41. For a reading that draws attention to the plant analogies in Wollstonecraft's work, see Alan Bewell, "'Jacobin Plants': Botany as Social Theory in the 1790s," also discussed in Anka Ryall's "A Vindication of Struggling Nature: Mary Wollstonecraft's Scandinavia." Bewell emphasizes the radical nature of Linnaean analogies between animal and plant reproduction, and makes a case for their impact on Wollstonecraft's use of plant analogies for women's development in *Vindication of the Rights of Woman*. (This is also a usage he identifies as deconstructive: having rejected those floral associations for women that would reduce their role to that of mere sweet flowers, Wollstonecraft then turns the tables and "applies horticultural theory to understand the present state of women," redeploying the plant analogies to "demystify the social production of female sexual difference," 137). In this respect Wollstonecraft says that women *are*, after all, like exotic flowers—in the sense that their ornamental features are similarly overcultivated at the expense of their robustness and their "fruitfulness" (138). Bewell locates a number of botanical analogies employed by Wollstonecraft for women, focusing on the parallels between forced methods of cultivation. Thus, we see Wollstonecraft ask what kind of "soil" women need, and speak against cultivating women as luxuriates or exotics or as "beautiful flaws of nature." The points Bewell also suggests as particularly consequential are, first, that Linnaeus located analogies in the plant world for the social relations between humans, but that, second, although the direction of "analogical traffic" involved a reading of human sexuality into the plant world, this contributed to a more general analogical "traffic" in both directions. For her part Ryall, also referencing this argument, finds references to Linnaean botany in Wollstonecraft's "Letters Written During a Short Residence." These readings do not, however, interrogate the additional analogical effect generated when Wollstonecraft is variously likening women to plants, animals (ornamental, uncultivated and domesticated), brutes, children, savages, *and* slaves.
42. "At a time when abolitionist opinion was rapidly gathering strength, the appeal to anti-slavery sentiments was a significant polemical move," (Taylor, *Mary Wollstonecraft*, 240). On the contemporary, negative connotations of the West Indian slave owner see Srividhya Swaminathan, "Developing the West Indian Proslavery Position after the Somerset Decision," in *Slavery and Abolition* 24, 3 (2003):40–60.

43. Taylor notes that for Wollstonecraft there is also a religious inflection to Wollstonecraft's affirmation of women's moral claims. Equiano's narrative makes a number of references to the ideals of Christian morality in its denunciation of slavery.
44. Mary Wollstonecraft, "The Interesting Narrative of the Life of Olaudah Equiano, or Gustavus Vassa, the African. Written by Himself, 2 vols. (Review, May, 1789, Article VI)," in *The Works of Mary Wollstonecraft*, ed. Janet Todd and Marilyn Butler (New York: New York University Press, 1989), vol. 7:100–101.
45. See Bewell, "'Jacobin Plants': Botany as Social Theory in the 1790s."
46. On the widespread use of the analogy (and particularly on the differences between the American and British contexts in which the analogy was deployed), see Clare Midgley, "British Abolition and Feminism in Transatlantic Perspective," in *Women's Rights and Transatlantic Antislavery in the Era of Emancipation*, ed. Kathryn Kish Sklar and James Brewer Stewart (New Haven: Yale University Press, 2007), 144–168, 129.
47. In the context of which it is helpful to recall her point that a wrong cultivated physiology produces mental inferiority, and her comment that "dependence of body naturally produces dependence of mind." Wollstonecraft, *A Vindication*, 115.
48. Ibid., 113.
49. Ibid., 107.
50. Ibid., 213.
51. Ibid., 117.
52. Ibid., 102.
53. Ibid., 87.
54. Ibid., 105.
55. Ibid., 161–162.
56. Ibid., 143.
57. Ibid., 105.
58. Ibid., 243. Wollstonecraft also makes a comparison between animal habits of nurturing the young and the dignity brought to the process by the human nurturer. The latter takes on the charge of forming the understanding and contributing rationally to the perfectibility of the child. Thus the parent who fails in this project is closer to the animal world: "The parent who sedulously endeavours to form the heart and enlarge the understanding of his child, has given that dignity to the discharge of a duty, common to the whole animal world, that only reason can give." See Wollstonecraft, *A Vindication*, 246.

59. Ibid., 268.
60. Ibid.
61. Ibid., 108. She also suggests that, historically, women have been subjugated because of man's greater force, a great physical force she will allow as one true sexual difference.
62. See Ferguson, who describes how "Wollstonecraft's eighty plus references to slavery divide into several categories and sub-sets" (Ferguson, "Mary Wollstonecraft," 16). These different categories are helpfully disaggregated by Ferguson in her commentary—Wollstonecraft does not do so.
63. Moreover, one can add an additional layer here: the abridged version of Buffon reviewed by Wollstonecraft for *Analytical Review* contains a number of references to the brutal treatment of slaves, who are "abused, buffeted, treated like brutes." But it also reverses the analogy, containing depictions of classes of animals put to work for humans (horses and camels) in conditions described, in all their associated mistreatment and cruelty, as those of slavery, *Buffon's Natural History, Abridged*, 65, 71 and 223.
64. Wollstonecraft, *A Vindication*, 240.
65. Ibid., 248.
66. Ibid., 235.
67. Ibid., 91.
68. Ibid., 235.
69. Ibid., 270.
70. Ibid., 47. Wollstonecraft is seemingly limiting her criticisms at this point to the treatment and status of women of a greater level of wealth than could pertain to the laboring class. She is not thinking of female servants or agricultural laborers when she decries the woman's status as decorative lapdog. Taylor's discussion of Wollstonecraft's unfinished late novel *Maria* (1798) emphasizes that it is distinctive in the context of Wollstonecraft's work both for its greater attention to the differences between the forms and conditions of subordination of women of different classes, and also for a communication of a greater degree of solidarity, common sympathy or possibility of collective identification among them, albeit with limitations.
71. Wollstonecraft, *A Vindication*, 202. Note also the variations between the cultural images of slavery, which rely on images of the West Indian slave owner, and those that rely on images either of women's slavery, or slavery more generally in Egypt, China, and those "Oriental" countries to which seraglios are attributed.

72. Ibid., 77. See Sara Figal's chapter, "The Caucasian Slave Race: Beautiful Circassians and the Hybrid Origin of European Identity," in this volume for a reminder of the eighteenth-century imaginary circulating around the Ottoman slave market, and the "oriental" enslavement of women (the common reference to the "seraglio"). As Figal writes, these vague references circulated at the same time as a more specific view had emerged that Circassians offered the model of perfection for European beauty, and concurrent with a racial theory that white Europeans might have originated in the mountains of Caucasus.
73. Consider also Taylor's discussion of the passage from *Maria* in which Maria says of her dissipated husband Venables, that it is "as if an ape had claimed kinship with me." As Taylor comments, this parallel has a redoubled complication because the apishness of Venables is also being analogized to those "squalid inhabitants of . . . the lanes and the back streets of the metropolis." Taylor, *Mary Wollstonecraft*, 239.
74. Ibid., 268.
75. Ibid., 67.
76. Compare to Macaulay, who does use the image of the barbarous to denounce the enslavement of women (see Ferguson, "Mary Wollstonecraft," 14), but not the metaphor of brutishness.
77. Wollstonecraft, *A Vindication*, 268.
78. Ibid.
79. Ibid., 269.
80. Ibid.
81. Ibid., 273.
82. Ibid., 286. The passage continues: "To laugh at them then, or satirize the follies of a being who is never to be allowed to act freely from the light of her own reason, is as absurd as cruel; for that they who are taught blindly to obey authority, will endeavor cunningly to elude it, is most natural and certain."
83. Ibid., 285.
84. Ibid.
85. Ibid. Italics mine.
86. So it would continue through a large number of women's rights defendants, including Mill, though with a number of variations, on this theme see again Sklar and Stewart, Introduction to *Women's Rights and Transatlantic Antislavery in the Era of Emancipation*, ed. Kathryn Kish Sklar and James Brewer Stewart (New Haven: Yale University Press, 2007), xi–xxiii.

87. However, not every association occurs. Contemporary European men are not deemed barbaric in their treatment of women, although the barbarous ages of mankind are described by Macaulay as periods in which women were enslaved. Macaulay does argue that women need education but not that men render themselves "like" barbaric savages in denying these claims.
88. With exceptions—slaves are not depicted as brutal in their treatment of their animals, for example.
89. On this issue in Wollstonecraft's work, see Ferguson, "Mary Wollstonecraft," 32–33.
90. To return to Taylor's discussion of this novel, *Maria* could be considered an exception given the common sympathy it expresses for, and closer attention to the wretched legal and economic circumstances both of its upper class heroine and for those of the lower class asylum attendant Jemima, see Taylor's *Mary Wollstonecraft and the Feminist Imagination*, 237–239.
91. I have not discussed the function of the plant analogies here, but to open a door in that direction, to some extent there is a parallel. For each of the other types of analogy, Wollstonecraft can argue that women are reduced to being "no better than": slaves, savages, servants of superficial tastes and abilities, children, domestic animals, and working animals. And Wollstonecraft does mention and reject the view of women as like (mere) pretty flowers. However, as seen in footnote xcii, plant analogies are also used more specifically in reference to the question of what environments and conditions plants (fruiting trees, flowering plants) need in order to flourish best, or by contrast, to evoke the poor practices of cultivation that produce what she calls the "beautiful flaws in nature" (*Vindication*, 107). Thus women, likened to unnaturally cultivated plants, are, again, by virtue of the analogy, associated with an aberrant status, but in this case it is not because they are reduced to being "no better than plants" but because they are like badly or unnaturally cultivated plants, in that they are not provided with the appropriate "soil" or conditions.

9

REPRODUCING DIFFERENCE

Race and Heredity from a *longue durée* Perspective

STAFFAN MÜLLER-WILLE

The classification of mankind into three or four major "races"—white, black, and yellow or red—is still very much alive, even in the high-tech contexts of today's genomics and systems biology.[1] For example, the International Haplotype Map Project initially studied human genomic variation based on four "population samples." The choice of these samples is revealing: for its pilot study, the HapMap Project looked at "samples from Nigeria (Yoruba), Japan, China and the U.S. (residents with ancestry from Northern and Western Europe [. . .])."[2] This choice was no doubt guided by the long-established classification of humans into four big "races" according to skin color, as it was originally proposed by Carl Linnaeus (1707–1778) in his *Systema naturae* of 1735.[3] The HapMap Project thus exhibits a curious mixture of archaic concepts and the latest tools of molecular biology.

Examples like this indicate a recent resurgence of racial categories in genomics, which many observers have found surprising and unsettling.[4] After all, it was preceded by a broad consensus—both among practitioners and commentators, and dating back to the so-called UNESCO Statement on Race from 1951—that the concept of race belongs to the past and has been thoroughly outdated by the combined efforts of mathematical population genetics and molecular biology. However, in the wake of the completion of the Human Genome Project and with projects like the Human Diversity Project, the HapMap Project, various national "biobank" projects, and a diversity of private and public initiatives, racial categories appear to have regained significance in recent years, inside and outside the biomedical sciences.[5] Racial distinctions are used as "proxies" in projects that try to map health disparities onto patterns of genomic variation; drug and life-style recommendations target racially defined groups, and genetic tests purport to determine ancestry in racial terms. Increasingly, close historical scrutiny also reveals that, throughout the post–WWII era, race was not only occasionally put back on the agenda through high-profile publications such as Richard J. Herrnstein's and Charles Murray's *The Bell Curve* (1994), but actually persisted as a distinct, though little-publicized thread in medical and population genetics research, especially in epidemiological contexts.[6]

What is it, then, about concepts of race—introduced in a patently ad hoc fashion by Linnaeus, and again and again denounced as primitive and untenable by prominent life-scientists in the course of their long history—that lets them persist, despite the rapid, conceptual and technological advances that biology has seen, especially in the twentieth century? In the following, I try to give an answer to this question based on results from a long-term project on the history of the concept of heredity.[7] In a nutshell, my answer amounts to the following: the concept of heredity, when it entered biology in the early nineteenth century, did not refer to the fixity of species or the age-old observation that "like engenders like." It was geared toward a much more specific phenomenon—namely, that of "heritable variation." From very early on, as I explain in the first two sections of this chapter, hereditary diseases on the one hand, and racial characteristics on the other, formed the paragon of hereditary phenomena. This is of great significance for the history of the human sciences, including anthropology and medicine. First of all, the juxtaposition of pathological and racial characters led to a conflation of the normal and pathological, or the natural and the accidental. Moreover, it was only through the focus on the resulting, oxymoronic causal constellation—consisting in the production of an original, individual deviation, whose effects were then regularly reproduced in offspring—that a space opened for a truly historical outlook in the life sciences, as epitomized in Charles Darwin's theory of evolution.[8] Heredity does not follow the logic of natural kinds, but that of historical events with lasting effects. In conclusion, I come back to my original question, making the point that concepts of human race persist to this day because they have been indelibly inscribed in the conceptual architecture that has supported, and continues to support, all attempts at describing and controlling human variation on a global scale since the early modern period.

HEREDITY AND DISEASE

"Heredity" originally only had a legal meaning in all European languages and was derived from the Latin *hereditas*, meaning inheritance or succession according to rules specified by law.[9] It was only around 1800 that heredity began to be used as a metaphor to address phenomena of organic reproduction. In the German-speaking world, Immanuel Kant (1724–1804) seems to have been the first to do so in his anthropological writings in the 1770s and 1780s, to which I return later. The *Oxford English Dictionary* lists Herbert Spencer's *Principles of Biology* from 1863 and Francis Galton's *Hereditary Genius* from 1869 as the earliest references for "heredity" in the modern, biological sense. Carlos López Beltrán has studied in detail how French physicians—psychiatrists or so-called alienists in particular— by the late eighteenth century started to discuss heredity and later disseminated the "philosophical and physiological" use of the term throughout Europe. López Beltrán also points to a peculiar linguistic shift that accompanied the spread of this parlance—namely, a shift from an adjectival (*héréditaire*) to a nominal use (*hérédité*), indicating the reification of the concept, or in López Beltrán's words, the establishment of a "structured set of meanings that outlined and unified an emerging biological conceptual space."[10]

That physicians played a crucial role in the initial shaping of the discourse of heredity is no coincidence. In fact, the dating of the emergence of this discourse around 1800 must admit one notable, but rather narrowly circumscribed exception within medicine. Since the late medieval period, physicians had sporadically referred to diseases that were restricted to particular families as "hereditary diseases." Admittedly, as Maaike van der Lugt has emphasized, such diseases played a minor role in scholastic medicine. The dominating doctrine of disease was that of humoral pathology, which defined diseases as disturbances in the balance of the four body humors: blood, phlegm, yellow bile, and black bile.[11] The constitution of a particular person was believed to result from the specific proportion of these four humors and according to the preponderance of this or that humor physicians distinguished between sanguine, phlegmatic, choleric, or melancholic temperaments. What we would call *environmental factors* today, by contrast, were summarized as the six "nonnaturals": light and air, nutrition, movement, sleep, excretions, and emotions. These were considered non-natural because they could be influenced by the physician or patient, for example, by keeping to a certain diet, or through blood-letting. In general, therefore, diseases were identified with the states of individual bodies that were elicited by a variety of incidentally or periodically recurring factors, rather than as entities that could be abstracted from their manifestations in individual bodies.[12]

Thus, within the humoral framework there was little room for a conception of diseases, let alone bodily properties in general, that in any literal sense could be seen as being passed down or transmitted from parent to offspring. When metaphors of inheritance were used in the late medieval and early modern period with reference to diseases that were observed in certain families only, it was therefore not so much the inheritance of mobile and alienable properties—money, for example—that people had in mind, but rather the passing on of landed property. Thus Jean Fernel (1497?–1558) maintained in his *Medicina* (1554) that a son is "as well inheritor of his [father's] infirmities as of his lands."[13] The problem was not to explain how properties were transmitted, but rather to explain how the causal agents that once had been involved in the generation of ancestors could remain active in the generation of their remote descendents.[14] This is how William Harvey (1578–1657) was to formulate this problem more generally in his *Exercises on the Generation of Animals* of 1651: "The knot therefore remains untied [. . .], namely: how the semen of . . . the cock forms a pullet from an egg . . ., especially when it is neither present in, nor in contact with, nor added to the egg."[15]

It is precisely this conundrum—a variation on the problem of action at a distance—which also preoccupied those few physicians who in the late medieval and early modern period specifically devoted themselves to hereditary diseases. The first to do so was Dino del Garbo (c. 1280–1327), who taught medicine according to Avicenna's *Canon medicinae* in nothern Italian universities in the early fourteenth century. Two manuscripts from around 1320 survive, which contain disputations that deal with the question "whether a disease which is in the father can become hereditary in the son (utrum aliquis morbus qui esset in patri posset hereditarius in filio)."[16] Avicenna had already distinguished hereditary from contagious and regional diseases at the beginning of the *Canon medicinae*, but without any further comment.[17] Starting from an analysis of the juridical concept of heredity, Dino proceeded to distinguish between truly hereditary (*morbi ex hereditate*) and connate diseases (*morbi ex generatione*). Connate diseases resulted from events that caused a change in the seed or embryo during conception or pregnancy. Dino argued that hereditary diseases, by contrast, had to be present already in the father to warrant the analogy with the transmission of worldly goods. A similar distinction was made slightly later in medical manuscripts by John of Gaddesden (c. 1280–1361), court physician of Edward II in England.[18]

Dino del Garbo's careful distinction remained largely ineffective, however, even if the phrase "hereditary disease" from then on appeared occasionally in European medical literature. Many authors applied the term to diseases that, according to Dino, should have been considered connate, such as leprosy, which many believed to arise from intercourse during menstruation. Humoral pathology continued to dominate medical thinking, and Dino had indeed indicated that hereditary diseases were in need of a different explanatory framework. Referring back to Aristotle's theory of generation, he maintained that hereditary diseases could only be explained by a permanent change in the "formative power" (*virtus formativa*) of the male seed.[19]

It took almost three centuries for the next attempt at a systematic account of hereditary diseases to appear on the scene. In 1605, Luis de Mercado (1532–1611)—professor of medicine at Vallodolid and court physician of Philipp II of Spain—published *De morbis hereditariis*.[20] With Dino del Garbo, Mercado assumed that it was a permanent change in the *vis formativa* of the seed that caused hereditary diseases. Unlike Dino, however, he followed Galen in assuming that both the male and the female produce seed that mix in generating the embryo. Hereditary diseases consisted for him in a changed "character (*character*)" of the body, which he described as "preternatural (*praeter naturam*)," as it differed in some respect from the ordinary. To account for such changes, Mercado supposed an interesting metaphor. Because they had to be due to changes in the "virtue of the seed (*vi seminis*) of the parents, grandparent or great-grandparents," it seemed as if "nature regulated the generation of individuals by some instrument (*instrumento*) in such a way that they produce individuals deformed by a similar defect (*eadem labe foedatos*)."[21] The expression *character* is borrowed from the Greek and does indeed refer to instruments used for brandishing domestic animals (and slaves!) or stamping money. Alongside this, Mercado also frequently used the term *sigillatio*, derived from *sigillum*, Latin for seal, which in theology denoted the indelible, but invisible nature of the sacraments.[22]

This model of inheritance allowed Mercado to distinguish hereditary diseases sharply from connate ones, and it should be integrated with solidist conceptions of disease in the late eighteenth century, which explained disease by a permanent lesion in the structure of the body.[23] Yet even if Mercado thus sometimes spoke of hereditary diseases as "untreatable," he saw some room for therapy. On the one hand, symptoms could be suppressed by adequate diet or treatment, and over the course of several generations such a treatment would eventually change the hereditary character. On the other hand, it was possible to compensate for a diseased character by combining it with a healthy one in generation. In this way, Mercado also explained the curious fact that hereditary diseases sometimes jumped one or several generations. Hereditary diseases could remain latent.[24]

Dino del Garbo's and Louis Mercado's discussions of hereditary diseases demonstrate that heredity first emerged as a subject at the periphery of medical discourse. Today it seems evident that every disease has a genetic component. For physicians and natural philosophers in the late medieval and early modern period, however, hereditary diseases were one special form of disease alongside others. The hereditary transmission of diseases appeared as a curiosity. Thus, Michel Montaigne (1533–1592) asked himself when he began to suffer from gallstones at forty-five years of age, just like his father had at the same age, what kind of "prodigy (*monstre*)" was hidden in the male seed so that it not only transmitted "impressions" of the bodily conformation, but also of the "thoughts and inclinations of our fathers."[25] After all, Montaigne also shared with his father a pronounced antipathy toward physicians. For Montaigne, heredity was not a generalizable, natural phenomenon but an example of the "miracles in obscurity" with which nature confronts humans on a daily basis. What was so intriguing about the gallstones was that they appeared at the same age in both father and son even though the elder had not yet developed this ailment when he generated his offspring. All other examples that Montaigne listed to illustrate heredity share the eccentric character of curiosities. Members of the Roman family Lepidus were often born with one eye covered by cartilage; according to ancient tradition, a Thebian tribe was distinguished by a lance-like birthmark; and, if one believes Aristotle, some of the Greek tribes practicing "women . . . in common" determined paternity by means of such bodily oddities.[26]

Heredity was thus not seen as instantiating a natural law, but quite on the contrary, it belonged to the realm of individual peculiarities and accidental aberrations. "All things are governed by law" is the conventional translation for the opening sentence of the Hippocratic tract *De genitura* ("On Seed").[27] Yet, it is worthwhile to consult its Renaissance Latin translation: "Law strengthens everything (*Lex quidem omnia corroborat*)," where "law (νόμος)," as the translator Girolamo Mercuriale (1530–1606) carefully noted in a comment, means customs, pasture, region, tribe (*instituta, pascua, regionem, classem*).[28] Law, in *De genitura*, did not refer to universal laws of nature, but to the persistence of local tradition and circumstance. The foundation for similarities between parents and offspring was thus provided by the fact that, as a rule, similar conditions prevail during procreation and development. Conversely, this meant that any deviation from the ordinary course of things would produce equally deviant results. The reproduction of similarity was thus as trivial as it was precarious, always remaining vulnerable to disturbances and transgressions.

HEREDITY AND HUMAN VARIETY

That inheritance was largely seen as something restricted to special circumstances in the early modern period can be seen from the fact that Carl Linnaeus (1707–1778), who was one of the first to propose a universal classification of mankind according to skin color, still felt compelled to underline that skin color should be seen on par with other variable characteristics, like stature or body weight, that clearly depended on environmental factors such as nutrition.[29] It was only in the course of the eighteenth century, and especially toward its end, that the peculiar behavior of heritable characteristics—the fact that they were transmitted without being influenced by external conditions—began to be seen as an instantiation of something akin to natural laws. Kant played an important role in advancing this perspective, and it is to his writings on race that I now want to turn.

For Kant, variation in bodily characteristics was not enough to constitute a racial difference. To the contrary, the differentiation of humanity into a set of interrelated races constituted for him a narrowly circumscribed, highly specific phenomenon. He clearly distinguished racial characteristics from species-specific traits on the one hand, which did not differ at all throughout a species and thus seemed to obey some constant law, and variable traits on the other hand, which either differed in accordance with changing environmental conditions or did not obey any obvious rule at all in their appearance among offspring. Only racial traits, according to Kant's definition, were traits that were invariably transmitted to offspring even under changed environmental conditions, and yet would regularly and predictably blend in hybrid offspring. European parents, to use Kant's favorite example, would continue to produce white children even when living in Africa, and Africans would continue to produce black children even when living in Europe, while both together produced children of an intermediate, brown skin color, again, regardless of the particular environment in which they were born.[30] Such a phenomenon, as Robert Bernasconi emphasizes in his contribution to this volume as well, undercut the distinction of specific forms and accidental peculiarities. In characterizing classes at a subspecific level, racial characteristics belonged to the individual peculiarities that interfered with the universality of species. Yet, these peculiarities were being reproduced infallibly, generation by generation, and thus seemed to be subject to the same kind of laws that governed the reproduction of species.

To account for this, Kant brought together natural law and contingent family history in his concept of *Vererbung*. The dispositions, or *Anlagen*, for hereditary traits were included from the very beginning in the organization of the original stock of ancestors from which all of humanity sprang, and were in this sense preformed and not acquired in a reaction to particular circumstances. Once these dispositions had been expressed as actual traits in reaction to a particular climate, however, they would be permanently and irrevocably transmitted.[31] Again, this behavior might seem to be nothing but a curiosity. But as Raphael Lagier has argued, the explanation of this behavior occupied a central place in Kant's overall philosophical project. It was able to resolve the major conundrum, clearly realized by Kant, that the supposedly universal moral and epistemological values that underwrote the Enlightenment were of a distinct geographic origin; in short, the Enlightenment was a distinctively "European" achievement, at least in the eyes of Europeans.

Kant explained this conundrum by maintaining that the equipment of the original human stock with *Anlagen* for adaptations to various climates, such as skin color, served the purpose of a universal geographic distribution of humankind. This "natural" distribution was overrun by a further process— the spread of civilization—which Kant saw as uncoupled from the former. The partition of the human species into four races, caused by their adaptation to distinct climates, was for him thus nothing but an accidental feature in relation to the universal spread of civilization—itself a destiny of the human species in as much as it amounted to the full realization of man's rational faculties.[32] Despite being accidental with respect to the destiny of mankind, the partition provided Kant with a rationalization, if not justification, of why certain races would either be overrun by civilization or reduced only to play a certain role in it. Depending on their presumed propensity toward work and reproduction, again a product of climate and soil, Americans were bound to go extinct, Africans to become enslaved, and Asians to be left behind by the process of civilization. Civilization and progress were entirely European affairs; neither essentially, nor necessarily, but by an accident of natural history that happened to place Europe and its (white) inhabitants at the center of the "centrifugal space of human identity," as Lagier puts it.[33] Racial hierarchies, in other words, were not simply a product of nature, but of the natural history of mankind, involving migrations and adaptations to particular climates.[34]

What were the sources of Kant's curious concept of race? The concept of human races was not a simple invention of eighteenth-century naturalists like Linnaeus, Georges-Louis Marie Leclerc, Comte de Buffon (1707–1788), or Johann Friedrich Blumenbach (1752–1840), who were the authors on whom Kant relied for his account. These naturalists, in their turn, relied on travelers' accounts, which reported a curious system of social stratification instituted in the Spanish and Portuguese overseas colonies: the so-called *castas*. This classification scheme originated from attempts to find a measure by which legal and social status could be allocated to the various sections of colonial society. It seemed to be primarily based on a classification of people according to skin color and, to a lesser degree, also on hair form and eye color. And children resulting from mixed marriages seemed to be positioned in this scheme by analogy to the simple mechanism of color mixing, implying processes of transmission and "blending" that connected traits of parents with traits of their offspring. During the eighteenth century, the system of *castas* found expression in a rich, pictorial genre in Latin America with pictures devised as sets arranged in serial or tabular form. Each of the individual pictures shows a mixed couple and its child, and each bears an inscription that states the components entering the mixture, that is, each parent's *casta*, and the result of the child's *casta*.[35]

Despite its rigid appearance, the *castas* system remained in constant flux throughout the early modern period, as witnessed by a rich proliferation of *castas* terms. In fact, it was not despite, but just because it was so rigidly based on an abstract classification according to color and on inheritance as an equally abstract mechanism, that the *castas* scheme could cope with this proliferation. The distinction according to colors—white, black, and brown—analytically defined the positions for all sorts of intermediate and more complicated cases. And the transmission and blending mechanism offered a unified explanation for their coming about, in so far as it could be regarded as operating independently of particular circumstances. To determine the *casta* of a person, it therefore sufficed to know the *castas* of his or her parents. Due to its analytic and quasi-mechanical character, the *castas* system could absorb a wealth of new phenomena while remaining stable in its basic outlines. It could therefore also account for the more capricious phenomena of heredity, like "regressions" or "throwbacks," as exemplified by a special caste in the system, the *torna atras* issuing from a Spaniard and an *albina*, that is, a white, blonde, and blue-eyed woman, which among its great-great-grandmothers had one black woman. Their child was usually depicted with a very dark skin color. The system of *castas*, as Renato Mazzolini put it, constituted "a vast field of 'pre-Mendelian' investigation."[36]

It was not intended to do so. In its original, local context with European colonies in the Americas, clothes and occupations played as much an important role in the assignment of *casta* as the possession of particular physical characteristics.[37] As Renato Mazzolini argues in this volume, it was the European commentator who paid particular attention to physical characteristics such as skin color. Once abstract, but purportedly universal classifications had been established on this basis, however, a curious process of accretion set in that brought other properties, including medical, cultural, and political ones, back into their fold. This is particularly evident in Linnaeus's successive presentations of human diversity. In the tenth edition of his *Systema naturae*, published in 1758, he did not only note skin color and other physical traits for each of his four races, but also medical temperament (i.e., Americans turn out to be "choleric," Europeans "sanguine," Asians "melancholic," and Africans "phlegmatic"), moral characteristics, preferred clothing, and form of government.[38]

HEREDITY AND THE STRUGGLE FOR LIFE

The accretion of facts of widely different kinds under the same classification of human races that I described at the end of the previous section is of great significance, as it resulted in a conflation of the normal and the pathological in hereditarian thinking. Although a "sanguine" temperament—and hence a disposition toward the development of certain diseases according to medical doctrines of the time—is declared normal for Europeans, it would be abnormal for them to be phlegmatic, which again is normal for Africans, or so Linnaeus maintains. "Blue eyes (*oculis caeruleis*)," on the other hand, suddenly stick out as a rather odd peculiarity of Europeans in Linnaeus's racial scheme.[39] This tendency toward a conflation of the normal and the pathological became a conspicuous element, often highlighted explicitly, of eighteenth-century theories of heredity. Thus Pierre-Louis Moreau de Maupertuis (1698–1759) wrote in his *Venus physique* of 1745 on the occasion of numerous observations of African albinos or "white Negroes (*Nègres-blancs*)": "Whether one takes this whiteness as a disease, or due to some accidental cause, it would always be a hereditary variety, which establishes itself or is effaced in the course of generations."[40]

Such conflations of the natural and the pathological in late eighteenth-century accounts of human variation infused a lasting element of historicity into subsequent attempts to write the natural history of mankind. In a similar manner as Kant, Maupertuis believed that nature contained the sources of human variations, but that it was "chance or art (*le hazard ou l'art*)" that shaped these into distinct races in analogy with breeders who "each year create some new species, and destroy those which are out of fashion."[41] "If the [white] Negroe, who is presently in Paris," he therefore speculated about his chosen subject, "found a *Négresse* here who is as white as he is, he would perhaps have only black children with her, because the number of generations may not have been enough to erase the color of their first ancestors. But if one engaged for several generations in giving white-negroe women to the descendants of this Negroe . . . such alliances would strengthen the race."[42] The albino was a rare deviation, something earlier generations of naturalists would have addressed as "preternatural" or "monstruous." Yet suitable measures, according to Maupertuis, could establish albinism as a new norm.[43]

This recourse to analogies with breeding technologies, such as selection and inbreeding, to explain patterns of physical distinctness has a long legacy as well, taking us back once more to the late medieval period, but again in a very narrowly circumscribed context. While Dino del Garbo wrote his disputations on hereditary diseases, European universities also witnessed a revival of ancient conceptions of "noble blood" in the context of polemical disputes about the nature and status of nobility. At the University of Paris, for example, a number of satirical disputations asked jokingly, if noblemen were characterized by long ears, just like certain dog races. This echoed the language of contemporary literature on the breeding of falcons, dogs, and horses, which frequently employed notions of "nobility," and also is one of the earliest sources of concepts of race.[44] It was again around 1800 that a more somber note was added to this discourse when historians began to connect it with notions of "struggle" and "war." As Michel Foucault emphasized, a number of historians began to describe the English Civil War and the French Revolution as instances of a perennial struggle of distinct human races.[45] Thus the French historian Augustin Thierry retraced the origins of the French Revolution to the conquest of Gallic territories and their population by Germanic tribes.[46] In a letter to Friedrich Engels, Karl Marx revealingly called Thierry "*le père* [the father] of class struggle."[47]

This is not the place to retrace these complex developments. I highlight, however, two things: first, quite in line with the original, juridical meaning of heredity, racial characters were seen as the product of a (political) will that contained an element of caprice. The breeder "corrects forms and varies colors, thus producing the Harlequin, the Mopse, etc.," Maupertuis mused, and asked himself, why the "sultans," who keep "women of all known species" in their "serails" do not likewise resolve themselves "to make new species."[48] This implies, secondly, that racial diversity and destiny were not so much understood as the outcome of a necessary, natural law, but rather as the outcome of a historical process involving ever shifting constellations of diverse motivations and forces.

These two motives came to the fore with utmost clarity in Charles Darwin's theory of evolution. One of the statements most often repeated in his *Origin of Species* (1859) is that there is no "fixed law of development."[49] What Darwin aimed at with this statement was to deny that there were laws governing the development of species. There was neither a law guaranteeing the fixity of species, as Carl Linnaeus had believed a century earlier, nor a law constituting progress toward forms of ever higher complexity, as in the theories of species transformation put forward by Jean-Baptiste Lamarck or Darwin's grandfather Erasmus. Yet one of the main principles of Charles Darwin's theory, the "principle of divergence of character," relied on the regular occurrence of heritable variations. So if there was no general law of development, there were at least laws—or "tendencies," as Darwin preferred to call them[50]—that governed the reproduction of individuals. Inheritance, in particular, was defined by Darwin in a crucially peculiar way. Relying on "Dr. Prosper Lucas' treatise, in two large volumes," as well as the experiences of breeders, Darwin defended inheritance against "doubts thrown on this principle by theoretical writers" in the following way:

> When a deviation appears not unfrequently, and we see it in the father and child, we cannot tell whether it may not be due to the same original cause acting on both; but when amongst individuals, apparently exposed to the same conditions, any very rare deviation . . . appears in the parent . . . and it reappears in the child, the mere doctrine of chances almost compels us to attribute its reappearance to inheritance.[51]

Inheritance, as is evident from this passage, was for Darwin not a process that simply accounted for similarities between parent and offspring. Such similarities could easily be, and had indeed been for a long time, explained by assuming that similar causes remained active in generation. Much more specifically, inheritance comprised cases in which a *difference* or *deviation* occurred, which was then reproduced despite the fact that both varieties lived under essentially the same conditions. "The saying that 'like begets like'," as Darwin stated in *Variation of Plants and Animals under Domestication* (1868), "has, in fact, arisen from the perfect confidence felt by breeders, that a superior or inferior animal will generally reproduce its kind."[52]

It is important to notice that inheritance thus defined, as well as its counterpart, variation, turn out to be capricious, not necessarily adaptive tendencies. "Inheritance operates fitfully," as Robert Bernasconi perceptively observes in this volume. Variation leads to differences reproduced under essentially the same environmental conditions, and by inheritance, identities are retained even if conditions change. In both cases, the results may, but certainly do not have to amount to perfect adaptation. The two sources that Darwin drew on, Prosper Lucas's (1805–1885) *Traité philosophique et physiologique de l'hérédité naturelle* (1847), which epitomizes a long medical tradition of dealing with hereditary diseases, and breeders' knowledge about the production of distinct and constant races, make this point abundantly clear. In both cases, inherited differences are conceived of as individual deviations from a norm, in the former case nonadaptive and in the latter case adaptive, if only measured against the strict selection regime imposed by the breeder. At the heart of living nature, and not only in its more exceptional productions, such as albinos, Darwin diagnoses capricious forces at work. Hereditary variation is clearly at odds, both with a view that sees organisms as always already adapted to their environments and a view that regards organisms as having limitless plasticity in their interaction with the environment. "Divergence of character," brought about by the combined effects of variation, inheritance, and natural selection, therefore can, but must not necessarily happen. It depends entirely on the degree to which a new variety is indeed better adapted to the prevailing conditions of life. "We must not . . . assume that divergence of character is a necessary contingency," as Darwin put it, "it depends solely on the descendants from a species being thus enabled to seize on many and different places in the economy of nature."[53]

With this perspective, nothing can be taken for granted in the long run. It is true that most nineteenth-century theories of heredity were directed toward the economically, socially, and politically disadvantaged, and thus exhibit what Carlos López Beltrán has termed the "hereditary bias" of dominant elites.[54] This bias should not be mistaken, however, as a sign for steadfast belief in a static, natural hierarchy of "primitive" and "advanced" forms of life. Characteristically, the ancient idea of a scale of beings saw different stages of perfection as necessary ingredients of nature. The modern idea of an evolutionary scale, by contrast, views the less perfect or "primitive" manifestations of nature as something that could, and usually would, eventually be overcome. By the same token, however, permanent success was not guaranteed. Ideologies of progress were therefore invariably coupled with deep-rooted fears of degeneration and calls for technologies of elimination and purification that could counteract the spontaneity of nature.[55]

CONCLUSION

As is well-known, Darwin's theory of natural selection was met with indifference, if not hostility, in the medical community if we leave aside the important exception of eugenics. Cell biologists like Rudolf Virchow, physiologists like Claude Bernard, bacteriologists like Louis Pasteur, all remained skeptical about this theory if they commented on evolution at all. This skepticism, however, should be seen as specifically directed against the principle of natural selection. Another centerpiece of Darwin's theorizing, the principle of inheritance—that which starts out as a rare deviation may come to constitute a new norm—did have a huge impact on biomedical thinking. It opened prospects for both the production of specific differences through targeted intervention and the reliable reproduction of such differences in the form of "purified" populations of controlled inheritance. Specific forms ceased to be eternal and immutable, and entered the realm of what could be created and manipulated. In other words, the recognition of heredity as a central life force opened the road for the entry of targeted, experimental interventions and well-defined model organisms in the life sciences.[56]

With respect to humans, experimental intervention and genetic manipulation will always remain problematic, if not anathema. The use of human subjects for experimental purposes is fraught with moral and political problems. I believe this is precisely why we see biologically crude, but historically, socially, and politically significant racial categories constantly reappear in biomedical research. Albeit ad hoc and "artificial," they provide the conceptual grid of last resort to gauge patterns and capacities for hereditary variation in humans, as the organizers of the HapMap Project readily concede.[57] In a sense, as physical anthropologists of the nineteenth and twentieth centuries have insisted on again and again, it is the history of mankind, with its alliances, migrations, revolutions, and conquests, which provides the human sciences with its one and only experiment.[58] This also means that a dismissal of the concept of human races as "essentialist" or "typological" will usually miss the point.[59] The science of heredity, and with it the concept of human races, has not left us more subjected to nature. On the contrary, it has turned our very nature into something that is historical through and through, and hence lies open to future projections.

NOTES

1. See Staffan Müller-Wille and Hans-Jörg Rheinberger, "Race and Genomics: Old Wine in New Bottles?" *NTM—Journal of the History of Science Technology and Medicine*, 16 (2008):363–386.
2. See "About the International HapMap Project," http://www.hapmap.org/abouthapmap.html, last access August 31, 2012.
3. Carl Linnaeus, *Systema naturae* (Leiden: Haak, 1735).
4. See for example Deborah A. Bolnick et al., "The Science and Business of Ancestry Testing," *Science* 318 (2007):399–400.
5. Barbara A. Koenig, Sandra Soo-Jin Lee, and Sarah S. Richardson, eds., *Revisiting Race in a Genomic Age* (New Brunswick: Rutgers University Press, 2008).
6. Claudio Pogliano, *L'Ossessione della razza: Antropologia e genetica nel XX secolo* (Pisa: Edizioni della Normale, 2005); Jenny Reardon, *Race to the Finish: Identity and Governance in an Age of Genomics* (Princeton: Princeton University Press, 2005); Keith Wailoo and Stephen Pemberton, *The Troubled Dream of Genetic Medicine: Ethnicity and Innovation in Tay-Sachs, Cystic Fibrosis, and Sickle Cell Disease* (Baltimore: Johns Hopkins University Press, 2006).
7. Staffan Müller-Wille and Hans-Jörg Rheinberger, eds., *Heredity Produced: At the Crossroads of Biology, Politics, and Culture, 1500–1870* (Cambridge, MA: MIT Press, 2007).
8. See Staffan Müller-Wille and Hans-Jörg Rheinberger, *A Cultural History of Heredity* (Chicago: University of Chicago Press, 2012), 72–75.
9. Franck Roumy, "La naissance de la notion canonique de *consanguinitas* et sa réception dans le droit civil," in *L'hérédité entre Moyen Âge et Époque moderne: Perspectives historiques*, ed. Maaike van der Lugt and Charles de Miramon (Firenze: Sismel—Edizione del Galluzzo, 2008), 41–66.
10. Carlos López Beltrán, "The Medical Origins of Heredity," in *Heredity*

Produced. At the Crossroads of Biology, Politics and Culture, 1500–1870, ed. Staffan Müller-Wille and Hans-Jörg Rheinberger (Cambridge, MA: MIT Press, 2007), 105–132; see also, Carlos López Beltrán, "In the Cradle of Heredity: French Physicians and *l'hérédité naturelle* in the Early Nineteenth Century," *Journal of the History of Biology* 37 (2004):39–72.

11. Maaike van der Lugt, "Les maladies héréditaires dans la pensée scolastique (XIIe-XVIe siècles)," in *L'hérédité entre Moyen Âge et Époque moderne: Perspectives historiques*, ed. Maaike van der Lugt and Charles de Miramon, (Firenze: Sismel—Edizione del Galluzzo, 2008), 41–66.
12. On humoral pathology, see Erich Schöner, *Das Viererschema in der antiken Humoralpathologie* (Wiesbaden: Steiner, 1964).
13. As translated in 1621 by Robert Burton, *The Anatomy of Melancholy* (1652), Project Gutenberg http://www.gutenberg.org/files/10800/10800-h/ampart1.html, last access October 20, 2012, 61. The Latin original has *haeredes* (in the plural) for "inheritor" and *possesionum* for "land"; see Jean François Fernel, *Medicina ad Henricum II Galliarum Regem Christianissimum, Vol. 2, Pathologia* (Paris: Andreas Wechel, 1554), 15.
14. Erna Lesky, *Die Zeugungs- und Vererbungslehren der Antike und ihr Nachwirken*, (Mainz: Franz Steiner, 1951; = Akademie der Wissenschaften und Literatur, Abhandlungen der Geistes- und Sozialwisssenschaftlichen Klasse, Jahrgang 1950, Nr. 19), 146–155.
15. William Harvey, "On Animal Generation," in *The Works of William Harvey*, ed. and trans. Robert Willis (London: Sydenham Society, 1847), 169–518, 354. On the ancient legacy of this problem, see Johannes Fritsche, "The Biological Precedents for Medieval Impetus Theory and its Aristotelian Character," *British Journal for the History of Science* 44 (2010):1–27.
16. Quoted from van der Lugt, "Les maladies héréditaires," 277, n16.
17. Ibid., 281.
18. Ibid., 288.
19. Ibid., 300.
20. The treatise first appeared in Valledolid in 1605 as part of the second volume of Mercado's collected works; see David F. Musto, "The Theory of Hereditary Disease of Luis Mercado, Chief Physician to the Spanish Habsburgs," *Bulletin of the History of Medicine*, 35 (1961):346–373, 346. I have only been able to consult the following, posthumous edition: Luis Mercado, "De morbis haereditariis," in *Lud: Mercati Operum Tomus II* (Frankfurt: Palthenius, 1608), 672–682.
21. Mercado, "De morbis haereditariis," 672; translation quoted from Musto,

"The Theory of Hereditary Disease of Luis Mercado," 361.
22. Mercado, "De morbis haereditariis," *passim*; see van der Lugt, "Les maladies héréditaires," 293, for a discussion.
23. López Beltrán, "The Medical Origins of Heredity."
24. Musto, "The Theory of Hereditary Disease of Luis Mercado," 361–362.
25. Michel de Montaigne, *Essais: Texte original de 1580 avec les variantes des éditions de 1582 et 1587*, ed. R. Dezeimeris and H. Barckhausen, 3 vols. (Bordeaux: Féret, 1873), vol. 2:332; translation quoted from Michel de Montaigne, *The Complete Essays of Montaigne*, trans. Donald M. Frame (Stanford: Stanford University Press, 1976), 578.
26. Montaigne, *The Complete Essays*, 578.
27. Hippocrates, *Hippocratic Writings*, ed. G.E.R. Lloyd (Harmondsworth: Penguin Books, 1950), 317; see Hans Stubbe, *History of Genetics: From Prehistoric Times to the Rediscovery of Mendel's Laws* (Cambridge, MA: MIT Press, 1972), 18–21, for a discussion of this important Hippocratic tract.
28. Hippocrates, "De genitura," in *Hippocratis Coi Opera qvae Graece et latine extant*, ed. and trans. G. Mercuriale (Venice: Industria ac sumptibus Iuntarum, 1588), vol. 1:10–16, 10 and 15.
29. Carl Linnaeus, *Critica botanica* (Leiden: Wishoff, 1737), 152–155.
30. Immanuel Kant, "Bestimmung des Begriffs einer Menschenrace (1785)," in *Kant's gesammelte Schriften, Abt. I.: Kant's Werke, Bd. VIII: Abhandlungen nach 1781* (Akademieausgabe), ed. Königlich Preußische Akademie der Wissenschaften (Berlin: Reimer, 1912), 89–106, 91–92; see Peter McLaughlin, "Kant on Heredity and Adaptation," in *Heredity Produced: At the Crossroads of Biology, Politics and Culture, 1500–1870*, ed. Staffan Müller-Wille and Hans-Jörg Rheinberger (Cambridge, MA: MIT Press, 2007), 277–291, for a discussion.
31. Kant, "Menschenrace," 99–100; see the essay by Roberto Bernasconi in this volume for a detailed discussion of Kant's theory of germs; for a classic account, see Timothy Lenoir, "Kant, Blumenbach, and Vital Materialism in German Biology," *Isis* 71 (1980):77–108.
32. Raphaël Lagier, *Les races humaines selon Kant* (Paris: Presses Universitaires de France, 2004), 186; see also Susan M. Shell, "Kant's Concept of a Human Race," in *The German Invention of Race*, ed. Sara Eigen and Mark J. Larrimore (Albany: State University of New York Press, 2006), 55–72, and Robert Bernasconi, *Nature, Culture, Race* (Huddinge: Södertörn University, 2010).
33. Lagier, *Les races humaines*, 193.

34. On Kant's concept of a history of nature, and its ambiguous legacy, see Phillip R. Sloan, "Kant on the History of Nature: The Ambiguous Heritage of the Critical Philosophy for Natural History," *Studies in History and Philosophy of Biological and Biomedical Sciences* 37 (2006):627–648.
35. Ilona Katzew, *Casta Painting* (New Haven: Yale University Press, 2004).
36. Renato Mazzolini, "Las Castas: Inter-Racial Crossing and Social Structure (1771–1835)," in *Heredity Produced: At the Crossroads of Biology, Politics and Culture, 1500–1870*, ed. Staffan Müller-Wille and Hans-Jörg Rheinberger (Cambridge, MA: MIT Press, 2007), 349–373, 365.
37. On this point, see chapter 5 in Ruth Hill, *Hierarchy, Commerce, and Fraud in Bourbon Spanish America. A Postal Inspector's Exposé* (Nashville: Vanderbilt University Press, 2005).
38. Carl Linnaeus, *Systema naturae*, 10th ed., 3 vols. (Stockholm: Salvius, 1758), vol. 2:20–21. Linnaeus's sources for these associations are unclear as he does not cite any. Traditionally, northern people were considered phlegmatic by humoral doctrine; see, Mary Floyd-Wilson, *English Ethnicity and Race in Early Modern Drama* (Cambridge: Cambridge University Press, 2003), 86. It is also interesting to see that Latin American physicians already began in the late sixteenth century to associate race with medical temperament, disagreeing widely about particulars; see, Carlos López Beltrán, "Hippocratic Bodies: Temperament and *Castas* in Spanish America (1570–1820)," *Journal of Spanish Cultural Studies* 8 (2007):253–289.
39. Linnaeus, *Systema naturae*, 10th ed., vol. 2:21.
40. Pierre-Louis Moreau de Maupertuis, *Vénus physique* (np: np, 1745), 161. My translation. The original reads: "Car qu'on prenne cette blancheur pour une maladie, ou pour tel accident qu'on voudra, ce ne sera jamais qu'une variété héréditaire, qui se confirme ou s'efface par une suite de générations."
41. Maupertuis, *Vénus physique*, 140–141.
42. Ibid., 139–140.
43. On literary tropes of beauty and breeding in late seventeenth- and eighteenth-century Europe see Sara Figal's contribution to this volume.
44. Maaike van der Lugt and Charles de Miramon, Introduction to *L'hérédité entre Moyen Age et époque moderne*, ed. Maaike van der Lugt and Charles de Miramon (Florence: Sismel—Edizione del Galluzzo, 2008), 3–37, 3.
45. Michel Foucault, *Society Must Be Defended: Lectures at the Collège de France, 1975-1976*, trans. David Macey (New York: Picador, 2003), 76–81 and 99–110.

46. Loïc Rignol, "Augustin Thierry et la politique de l'histoire: Genèse et principes d'un système de pensée," *Revue d'histoire du XIXe siècle*, 25 (2002), http://rh19.revues.org/document423.html, last access October 20, 2012.
47. Karl Marx and Friedrich Engels, *Collected Works*, vol. 3.7 (London: Lawrence & Wishart, 1975), 130.
48. Maupertuis, *Venus physique*, 141.
49. For example, Charles Darwin, *On the Origin of Species by Means of Natural Selection, or the Preservation of Favoured Races in the Struggle for Life* (London: John Murray, 1859), 314, http://darwin-online.org.uk/contents.html.
50. Darwin, *Origin*, 118.
51. Ibid., 12.
52. Charles Darwin, *The Variation of Animals and Plants under Domestication*, vol. 2 (London: John Murray, 1868), 2, http://darwin-online.org.uk/contents.html, last access October 20, 2012.
53. Darwin, *Origin*, 331.
54. Carlos López Beltran, *El sesgo hereditario: Ámbitos históricos del concepto de herencia biológica*, (Mexico City: Universidad Nacional Autónoma de México, 2004).
55. Anne Carol, *Histoire de l'eugénisme en France: Les médecins et la procréation, XIXe-XXe siècle* (Paris: Seuil, 1995).
56. Müller-Wille and Rheinberger, *Cultural History of Heredity*, chapter 6.
57. Richard A. Gibbs et al., "The International HapMap Project," *Nature* 426 (2003):789–796, 791; for the philosophical implications of this strategy, see Lisa Gannett, "The Normal Genome in Twentieth-century Evolutionary Thought," *Studies in the History and Philosophy of Biology and the Biomedical Sciences* 34 (2003):134–185.
58. On the history of this conception, see "BioHistories," Special issue, ed. Soraya de Chadarevian, *BioSocieties* 5 (2010).
59. Staffan Müller-Wille, "Making sense of essentialism," *Critical Quarterly* 53 (2011):61–77. Nor does the adoption of population thinking, as Lisa Gannett has warned us, provide a bulwark against racism; see Lisa Gannett, "Racism and Human Genome Diversity Research: The Ethical Limits of 'Population Thinking,'" *Philosophy of Science* 68 (2001):479–492.

10

HEREDITY AND HYBRIDITY IN THE NATURAL HISTORY OF KANT, GIRTANNER, AND SCHELLING DURING THE 1790s

ROBERT BERNASCONI

THE ROLE OF THE *KEIME* IN KANT'S THEORY OF RACE

This essay addresses an apparent puzzle in the history of science: Why did certain leading intellectuals of the 1790s adopt what can broadly be called the Kantian notion of race, when that notion was still largely being presented in the apparently discredited terms of *germs* or *seeds* (*Keime*)? The puzzle arises because commentators tell us that scientists had abandoned the *Keime* with the victory of epigenesis over preformationism as theory of generation. But whereas it is true that epigenesis had come to dominate, the assumption of these commentators is that all uses of the term *Keime* are incompatible with epigenesis. Indeed, some of these same commentators argue that Kant was forced to abandon the *Keime*. I argue that the reports of the debates from the late eighteenth century have been simplified to the point of distortion in the secondary literature.[1] This was possible, in part, because of a tendency to overschematize the debates, but also because, until recently, some historians of science of this period chose to ignore the discussions of race, which can now be seen as central to the discussion.[2]

I show that although Blumenbach, through his notion of the *Bildungstrieb*, gave a strong boost to epigenesis and ruled out some narrow uses of the word *Keime* within certain preformationist theories of reproduction, this account did not exclude reference to *Keime* in other contexts, including discussions of race, because the *Keime* were appealed to in another sense and for another purpose. What most clearly establishes that Kant's early discussion of race is not refuted by Blumenbach's *Bildungstrieb*, even though the former relies on a certain notion of *Keime* and the latter is directed against *another* notion of *Keime*, is the role of hybridity in both accounts. Commentators have tended to assume that the widespread adoption of Blumenbach's notion of *Bildungstrieb* had somehow discredited all uses of the notion of *Keime*, which, alongside that of *Anlagen*—in the sense of inclinations or predispositions—was the main tool by which Kant tried to make sense of how monogenesis could be reconciled with the idea of permanent racial divisions. However, Pauline Kleingeld recently acknowledged that Kant did not drop reference to the *Keime* and that more research was needed in this area.[3] The present study, which includes an overview of the reception of Kant's notion of race, begins this task.[4]

Let me briefly recall the problem that led Kant to turn to the language of *Keime* and *Anlagen* in giving his account of the different human races in the first place. In 1775, when Kant wrote his first essay on race (subsequently expanded in 1777), the dominant scientific view about the source of the varietal differences among human beings highlighted the impact of the environment, but more and more evidence was emerging concerning the permanence of certain inheritable characteristics. More precisely, the upholders of the unity of the human species had relied on a straightforward environmental explanation to account for the varietal differences, but the lack of any evidence of the kind that Buffon had anticipated of blacks becoming white and whites becoming black by virtue of moving to an area with a different climate rendered this claim unsustainable.[5] This put pressure on the monogenists of the day, who, in order to uphold the biblical account of the unity of the human species, usually maintained that all humanity must have derived from a single pair. As a result, they needed an explanation of how permanent differences could arise from the same source.

Such an explanation is precisely what Kant sought to supply with his account of human races in terms of *Keime* and *Anlagen*. Kant defined the *Keime* in the 1775 essay, in a passage repeated in the 1777 version, as "the grounds of a determinate unfolding which are lying in the nature of organic body (plant or animal)" in the case when that unfolding concerns particular parts. When the unfolding concerns the relation of the parts to each other, then the grounds are called "natural predispositions (*Anlagen*)."[6] The basic idea was that the capacity to adapt to any given climate must have been in the original human beings, but as their immediate descendants adapted to the environment in which they found themselves, they lost the capacity to adjust to other environments. That is to say, already in 1775, Kant's focus fell on purposive adaptations to the environment that, once acquired, became permanent.[7] The original human beings must have had the potential to become any one of the four races, but to the extent that those characteristics had been realized, the other possibilities were no longer available to their progeny, and there could be no reversion to the original type. The *Keime* and *Anlagen* were introduced by Kant as his explanation of how specifically racial characteristics are to be understood as inherited and permanent, given that the different races all belong to the same species and so descend from the same stem.

Kant had long shown a greater interest in heredity than many of his contemporaries.[8] In his 1763 essay "The Only Possible Argument in Support of a Demonstration of the Existence of God"—in the course of announcing his interest in what we now know as heredity—he expressed his dissatisfaction with the theories of Buffon and Maupertuis. The novelty of Kant's interest is evident in the lack of an available language with which Kant could pose the problem, as is evident from the following formulation:

> The purpose of these considerations has simply been to show that one must concede to the things of nature a possibility, greater than that which is commonly conceded, of producing their successors (*Folgen*) in accordance with universal laws.[9]

For Kant the question of heredity (*Vererbung*) moves from being focused largely on the transmission or bequeathment of hereditary diseases to the question of the capacity of an organism to adapt to its environment, according to ways which were set in advance or preprogrammed without being preformed.[10] This helps to explain the role that the *Keime* played for Kant: it was his speculative way of explaining the permanence of racial characteristics, that is, their hereditary character.

For the most part, inheritance operates fitfully and thus comes to be understood as contingent, but Kant defined racial characteristics as permanent, and this gave to race—skin color and all the moral and intellectual characteristics that indicated to Kant—a preeminence that it had hitherto lacked.[11] Indeed, much of the discussion of heredity for decades to come was intertwined with the discussion of race. Kant did not say explicitly that race is destiny, but that was in effect his position once the original adaptations had been made. According to Kant, there were four primary races corresponding to the original potential of the *Keime*. The races were formed in response to the climate in which they found themselves as a result of their initial migration. Varieties of human beings that did not fall neatly into the divisions between the four races did not refute it, but were evidence of an incomplete transformation, perhaps, because a people migrated again while their initial adaptation was still incomplete. Now only race-mixing allowed for further transformations.

That Kant's account of race concerns inheritance is most clearly announced in 1785 in a second essay on race, whose title could be translated as "Definition (*Bestimmung*) of the Concept of a Human Race." Kant offered there the following definition of race: "The classificatory difference of one and the same phylum (*Stamm*) in so far as this difference is unfailingly hereditary."[12] For Kant, the real test of inheritance was what he called "The Law of Necessary Half-Breed Generation."[13] The idea, as Kant had most clearly explained in the 1775 essay, was that one can clearly see the operation of inheritance in the case of race-mixing. It formed part of the definition of race that he gave there:

> Among the subspecies, i.e., the hereditary differences of the animals which belong to a single phylum, those which persistently preserve themselves in all transplantings . . . over prolonged generations themselves and which also always beget half-breed young in the mixing with other variations of the same phylum are called *races*.[14]

In other words, in clear distinction from the dominant debate about generation, which argued about whether the child was a product of the sperm or the egg, Kant—focusing on the inheritance of racial characteristics—posited that both parents played an equal role. He highlighted the mulatto, whose brown skin color was purportedly halfway between that of a black and a white parent.

According to the definition of race Kant supplied in his 1775 essay cited above, racial characteristics are not simply inherited characteristics, but are restricted to those inheritable characteristics in the progeny that are invariably midway between that of the parents, in such cases where the parents belong to two distinct races. In Kant's example, because a child has a skin color halfway between that of the parents, one knows that the child's parents belong to two distinct races.[15] Kant made the effects of race-mixing central to the identification of specifically racial characteristics and thus the identification of the main races themselves. Furthermore, it was the lack of consistency in most forms of inheritance—the fact that some characteristics may skip a generation—that brings racial inheritance to the fore. In other words, only if one allows oneself to look across several generations can one identify racial characteristics. For Kant, race is not open to simple description. It is not a surface phenomenon. It requires a special kind of investigation, which Kant calls "natural history," to identify racial characteristics as such.

At the time when Kant was initially formulating his racial theory, preformationism was dominant, and Kant employed the language of preformationism for his own purposes. But this did not set him in opposition to epigenesis. Theories of epigenesis had long accommodated the notion of *Keime*. One sees this, for example, in Johann Nicholas Tetens's discussion of the perfectibility and development of human beings that appeared in his 1777 book *Philosophische Versuche über die menschliche Natur und ihre Entwicklung* (Philosophical essays on human nature and its development), a work known to have had great significance for Kant.[16] It is true that Kant, following Haller's terminology, described the transformations brought about by migrations as preformed (*vorgebildet*) in the germs (*Keime*) and endowments or predispositions (*Anlagen*) that nature or providence had supplied in its solicitude.[17] What Kant was acknowledging in 1775 was simply "the mere faculty to propagate its adopted character."[18] But he needed more than what he called *Zeugungskraft*, or the power of reproduction, to explain their operation.[19] This was part of Kant's attraction to Blumenbach's account of the *Bildungstrieb*, which he already embraced in 1788 in his third essay on race, "On the Use of Teleological Principles of Philosophy," where he continued to employ the language of *Keime*.[20] The different seeds were described here as "originally implanted in one and the same phylum and subsequently develop *purposively for the first general population*."[21] The context of this statement was Kant's observation that Forster agreed with him that "Negroes" had hereditary characteristics fundamentally different from those that defined "Whites." That was not where the disagreement lay. Kant appealed to the *Keime* to maintain his support for monogenesis against Forster's advocacy of polygenesis.

BLUMENBACH'S REJECTION OF THE *KEIME* IN HIS ESSAY ON THE *BILDUNGSTRIEB*

In 1789 in *Über den Bildungstrieb* (An essay on generation), which was a completely different essay from *Über den Bildungstrieb und das Zeugungsgeschäfte* (On the formative drive and the business of procreation) published in

1781, Blumenbach rejected any suggestion that preexisting *Keime* could serve as the chief principle of generation, growth, nutrition, and reproduction:

> That there is no such thing in nature as pre-existing germs: but that the unorganized matter of generation, after being duly prepared, and having arrived at its place of destination takes on a particular action or *nisus*, which *nisus* continues to act through the whole life of the animal, and that by it the first form of the animal, or plant is not only determined, but afterwards preserved, and when designed, is again restored.[22]

In this essay, Blumenbach highlighted the generation of "bastards" and an experiment by Koelreuter in which he was able to turn one species of tobacco into another by crossing tobacco plants. Koelreuter was able to create "a complete metamorphosis of one natural species of plants into another."[23] Indeed, in the 1781 edition of *Über den Bildungstrieb* Blumenbach announced that the existence of mulattoes contradicted "all concepts of preformed germs."[24] Clearly Kant's use of *Keime* in his essays on race could not be referring to preformed *Keime* in that sense because the existence of the mulatto was the cornerstone of Kant's racial theory.

When Blumenbach rejected preexisting germs, he was rejecting those forms of preformationism that explained reproduction in such a way that the characteristics of the child derived from only one of the parents, depending on whether the sperm or, as with Haller and Bonnet, the egg was indicated.[25] Kant understood this point very well and reiterated it when, in 1790, in section 81 of the third critique he took up Blumenbach's account in order to dismiss individual preformationism as a prelude to offering his full support for Blumenbach's *Bildungstrieb*.[26] In its place he proposed what he called "generic preformationism" with the following explanation:

> There was one thing they [the individual preformationists] simply could not fit into their system of preformation: the production of bastards. In cases where the whole product is produced by two creatures of the same species, they had granted neither of these two a formative force, granting the seed of male creatures nothing but the mechanical qualification to serve as the embryo's first food. In the case of hybrids, however, they had to grant the male seed a purposively formative force as well.[27]

Blumenbach's 1789 essay on the *Bildungstrieb* provided the model that Kant himself used for setting out the rival theories of generation in the *Critique of Judgment*. Kant's source was easily recognized. When Salomon Maimon gave his account of the *Weltseele* in 1791, he rehearsed the different theories of generation with explicit reference to both Kant's *Critique of Judgment* and Blumenbach's account of epigenesis, and he also highlighted the account of the existence of bastards and organic parts that have never previously existed as a basis for rejecting the existence of what he judiciously called "preformed (*präformierte*) *Keime*."[28] It is possible that Blumenbach thought that Kant's use of *Keime* in his racial theory was irreconcilable with his own notion of *Bildungstrieb*, but if so, he never said it explicitly.[29] But for Kant, Maimon, and, as I go on to show, Girtanner and Schelling, there was no such problem and no need to abandon all *Keime*.[30]

The Vigilantius manuscript notes of Kant's lecture on metaphysics show the precise nature of the account of *Keime* that Kant rejected in 1790 in renouncing individual preformationism in the *Critique of Judgment* on the basis of hybridity. Kant associated the theory of *Keime*, which he rejected with the system of involution or encasement, which posited such that one germ would contain all the others as dead matter and that each of these would become animate only singly. Kant observed that an infinite sum of germs would have to be assumed in each individual, and he objected that all those possibilities would be wasted whenever a seed was isolated without propagating. However, in describing the system of epigenesis he emphasized that "the young animal . . . arises from the mixture of both sexes as a product." He continued:

> This is more likely, as already indicated by the mating of related kinds, e.g., donkey and horse, black and white human beings, the similarities of variations, in mules and mulattos, etc., etc., and so likewise the bastard plants produced by related pollen.[31]

It is again clear that to refute a certain theory of *Keime*, he appealed to his racial theory, a theory that relied on *Keime* in another sense.

To some commentators, Kant's phrase "generic preformationism" is evidence of confusion on Kant's part. Phillip Sloan has been particularly insistent that Kant's account of generation in section 81 of the *Critique of Judgment*, "On Conjoining Mechanism to the Teleological Principles in Explaining Natural Purposes as Natural Products," is idiosyncratic. He believes that to make room for Blumenbach's *Bildungstrieb*, Kant was forced to abandon his preformationist theory and that this is reflected between 1788 and 1790 in a shift from the model of *Keime* and *Anlagen* to that of "*zweckmässige Anlagen,*" which Sloan understands as phenomenological teleological forces distinct from preformed *Anlagen*.[32] More recently, in 2006, Sloan accused Kant of engaging in a sleight of hand to conceal the inadequacy of his own earlier accounts so as to integrate Blumenbach's attack on preformationism.[33] But Kant did not refer to epigenesis as "generic preformationism" simply out of confusion; still less to conceal his own change of view. He believed that epigenesis and a certain preformationism needed each other.[34] A certain epigenesis explained how organisms adapted to any new conditions, whereas a certain preformationism set the appropriate limits on that adaptability. Hence, Kant wrote in the third critique: "Nothing is to be taken up into the generative force that does not already belong to one of the being's undeveloped predispositions (*Anlagen*)."[35] This was a reiteration of the position held in the first essay on race.[36] Kant did not employ the word *Keime* in the *Critique of Judgment*, which has led to some speculation.[37] But it is hardly surprising that Kant avoided using the word, given the controversy concerning *Keime* and that he was not promoting the concept of race in that work although it is clear from the discussion of the *Anlagen* that one function of the third critique was to provide the underlying theoretical basis for his racial theory. In any event, Kant did not abandon the notion of *Keime*. In the lectures on physical geography delivered in 1791, as reported in 1831 by Friedrich Christian Starke (a pseudonym for Johann Adam Bergk), Kant uses the term *Keime* much as he had used it more than fifteen years earlier.[38]

These arguments show that the *Critique of Judgment* is not the watershed for the *Keime* that many commentators now assume it to be. To be sure, it seems unlikely that Kant would have thought that he could have switched sides on one of the most important scientific debates of his day without anyone noticing. In the last two sections of this chapter I show that if anyone believes that Kant was confused on this issue, then they must say the same of Girtanner and Schelling, among others. I believe that the confusion is actually in the secondary literature and is the result of a tendency to want to assign thinkers to fixed positions according to a set schema. For the most part, neither Kant, nor his most distinguished contemporaries, were trying to salvage dogmatically held positions at all costs. Rather, they were searching for answers to questions that they were still not sure how to ask. It would be extremely bizarre to suppose that Blumenbach's argument in favor of the *Bildungstrieb* refutes Kant's racial theory because his whole account of the inheritance of racial characteristics was dependent on the fact that he gave an equal role to both parents. To believe that Kant was forced by the example of hybridity to abandon the way he had framed his concept of race in 1775 is to believe that Kant had failed to recognize that the terms in which he couched his theory would not account for the precise case to which he appealed in order to establish this theory.

KANT'S ENDORSEMENT OF GIRTANNER'S CONTRIBUTION

It is clear that Kant and Blumenbach ultimately had different interests and that this extends beyond the different emphases of their discussions of race. Blumenbach did not display a deep appreciation of the theoretical framework offered by Kant's third critique, even though Kant cited Blumenbach at one stage as an example of what he had in mind.[39] Nevertheless, in 1797, when Blumenbach finally offered his own definition of the word *race* as distinct from variety, what he said echoed and indeed referenced Kant:

> Race in its precise sense is a character produced by degeneration that operates through inevitable and necessary inheritance, as, for example, when Whites with Blacks produce mulattoes, or when Whites with American Indians produce mestizoes.[40]

In this formulation Blumenbach showed that he understood the importance of hybridity—the bastards of *Über den Bildungstrieb*—for Kant's understanding of race. Nevertheless, that does not mean Blumenbach understood what Kant meant by "natural history." They were at cross purposes. Kant believed that Blumenbach's notion of *Bildungstrieb* was a contribution to what he called *natural history*, but Blumenbach seems never to have embraced natural history in Kant's sense and especially not in his writings on human varieties.

Christoph Girtanner seems to have had a better understanding of Kant than Blumenbach, albeit not a perfect one. In his *Über das Kantische Prinzip für die Naturgeschichte* (On the Kantian principle for the history of nature), Girtanner united Kant's *Keime* and Blumenbach's *Bildungstrieb*. He wrote:

> Natural history, in the philosophical sense, divides the organized bodies into lines of descent according to their relationships with respect to generative organization. They are based on the common law of reproduction. The unity of the species comes from the unity of the

generative power. In this way, a natural system for the understanding comes into being, an arrangement of organized bodies under laws, and, to be sure, especially under the law of the formative drive.[41]

The "Kantian principle" referred to in Girtanner's title concerns race.[42] It is drawn from Kant's 1785 essay on race and is described as "the law of half-breed generation and the invariable transmission of everything that distinguishes the real (*wirkliche*) race."[43]

Girtanner's reading of Kant cannot be summarily dismissed for a simple reason: two years after its publication in 1796, Kant in his *Anthropology from a Pragmatic Point of View* recommended it, which was for him an unusual gesture. The section "On the Character of Races" begins thus: "With regard to this subject I can refer to what Herr Privy Councilor Girtanner has presented so beautifully and thoroughly in explanation and further development in his work (in accordance with my principles)."[44] If Kant was intent on withdrawing his commitment to the notions of race, *Keime*, or natural history, he could have easily avoided this celebration of a work, which, in his name, took up all three notions. The second edition of the *Anthropology* in 1800 gave Kant a chance to remove his endorsement of Girtanner, but he did not take it.[45] It is also significant that, when referring to *Über das Kantische Prinzip für die Naturgeschichte*, Blumenbach no more raised a problem about Girtanner's continued references to the *Keime* than Kant did.

Girtanner was a significant author in his own right.[46] In *Über das Kantische Prinzip* he did more than record Kant's account of race and of natural history. He employed his extensive knowledge to expand its scope. But, above all, he made the account experiment-based. Whereas Kant highlighted the speculative character of natural history, Girtanner's focus was on what could be known with certainty, which, following the application of Kant's principle of half-breed generation, meant that "there is . . . nothing irresolute, indefinite or uncertain here."[47] One sees this in the way he joined Kant in shifting the understanding of heredity from focusing on monstrosities and diseases to understanding it in terms of races. In other words, the task of a science of nature was less to explain the exception than to identify the rule.[48] Even more striking is the way Girtanner handles Kant's reliance on children, whose skin color was halfway between that of their parents, to identify those parents as of different races.[49] This is a point on which Girtanner separated from Blumenbach. Although Blumenbach understood the significance of interbreeding across races for natural history, he did not see that it could form the basis of an experimental method. Indeed in 1795, in the third edition of *De generis humani varietate nativa*, Blumenbach had complained that Buffon's law, which used hybridity to identify species, was not particularly helpful to scientists because they would have to wait for intermixing to happen before they could determine answers to the questions:

> What very little chance is there of bringing so many wild animals, especially the exotic ones, about which it is of the greatest possible interest for us to know whether they are to be considered mere varieties, or as different species, to that test of copulation? Especially if their native countries are widely apart.[50]

When it came to identifying races, would one have to force people to have sex to see how their children looked before one could tell they were distinct races? This was especially ironic because Kant was against race-mixing.[51] It is no wonder that Blumenbach preferred to collect skulls. Kant's underwriting of Girtanner's book is clearly an obstacle to Sloan's thesis as it is to anyone who wants to claim that for some reason Kant abandoned his commitment to race and the *Keime*. Sloan insists that Girtanner's treatment of the issue reveals "uncertainty . . . about where he ultimately stood,"[52] but there is only uncertainty if one has already decided that the two terms—*Keime* and *Bildungstrieb*—cannot coexist.

SCHELLING'S TESTIMONY

Girtanner has played a role in the recent debate about Kant's alleged rejection of the *Keime*, in which he framed his account of race, but another important witness has not been called to testify: Friedrich Wilhelm Joseph Schelling. The importance of Kant's *Critique of Judgment* for the following generation of philosophers has long been recognized, but it is not always appreciated that their understanding of that work was to some extent mediated by their acquaintance with Kant's writings on race. This is especially clear in Schelling's *First Outline of a System of the Philosophy of Nature*. And of particular significance for the topic of this chapter, it showed that he clearly understood the need for Kant's notion of race to be combined with Blumenbach's *Bildungstrieb*. Schelling surmised that in natural history one must assume that in the first individuals of each species the direction of the *Bildungstrieb* was not yet set. Only on this condition could one say that these individuals were free to express their genus. And at this point, Schelling quoted—or, more precisely, misquoted insofar as there was a minor difference from the original text—the closing sentences of Kant's 1785 essay:

> It is impossible to guess how the form (*Gestalt*) of the first human phylum (*Stamm*) was constituted in respect of its color; even the character of the whites is only the development of one of the original predispositions that together with the others were found in the phylum.[53]

The point seems to be that the first individual of each species must have in some way contained the subsequent variations within the species without being able to express them directly.

Here Kant was trying to explain why the original stem or phylum from which the races had developed could never be restored, which is why he could insist at the end of the essay that we cannot now tell what the original was like. To explain the irreversibility of these changes Kant appealed to the germs: "That the germs which were originally placed in the phylum of the human species for the generation of the races must have developed already in most ancient times according to the needs of the climate, if the residence there lasted a long time; and after one of these predispositions was developed in a people, it extinguished all the others entirely."[54] At this point, Schelling directly embraced the language of Kant's account:

> Now that which is *developed* (but not, on that account, *brought forth*) through external influence is called *germ* or *natural predisposition*. Those determinations of the formative drive within the sphere of the general concept of the species, therefore, are able to be presented as *original natural predispositions* or *germs*, which were all united in the primal individual—but such that the prior development of the one makes the development of the other impossible.[55]

Schelling here adopted what was most innovative in Kant's notion of race: racial characteristics are permanently acquired such that once a race is formed, further changes are excluded. In other words, Schelling's answer to the question of what gave the *Bildungstrieb* its direction was, following Kant, "germs or natural predispositions."[56] Schelling thus provided the argument that Blumenbach's notion of the *Bildungstrieb* needed Kant's germs, not just in the case of race, but more generally. Schelling, who could easily have made the point by talking only of the *Anlagen*, specifically embraced the *Keime*; he took up both, and yet he was equally clear in his support of epigenesis. Schelling, in his *First Outline of a System of the Philosophy of Nature*, insisted that "all formation occurs through epigenesis."[57] This is a conclusion he arrived at on the basis of "natural history (in the authentic sense of the word)." And in manuscript notes, which are now included in the ongoing critical edition of Schelling's works, Schelling attributes to Blumenbach the notion of preformationism that Kant had identified in the third critique as his own: "We are agreed with Blumenbach in that there is no individual preformation in organic Nature but only a generic kind. Agreed, that there is no mechanical evolution, but only a dynamical one, and thus that there is only a dynamical preformation."[58] Nothing could be more distant from the picture that Blumenbach, by dismissing *Keime*, had forced Kant to renounce the terms in which he formulated his early discussion of race.

KANT'S LEGACY

During the last decade of the eighteenth century, the word *race* became more widespread, and it was frequently associated with Kant whenever there was a question of giving it a precise sense. The reservations about the scientific status of Kant's conception of race that were voiced by Johann Metzger, Georg Forster, and Johann Gottfried Herder in the 1780s seem to have been forgotten by the following generation of the 1790s. Interest in Kant's theory of race was not limited to theorists like Blumenbach, Girtanner, and Schelling. Lavater included extracts from Kant's 1777 essay on race in his *Essays on Physiognomy*. In this context it is worth noting that Lavater included some of the passages in which Kant referred to "the germs and propensities."[59] But perhaps the best indication of both the uniqueness of Kant's theory of race and the general public's interest in it is the space devoted to it in dictionaries dedicated to the task of explaining Kant's notoriously difficult terminology.

The Kantian lexica show that, at that time, the word *race* had come to be seen as Kant's word—a technical term that required explanation.[60] When Georg Mellin produced his massive dictionary of Kant's terminology, he devoted some seventeen pages to the term *race*.[61] Furthermore, Mellin highlights the role of both the *Keime* and *Anlagen* in this account.[62] Perhaps even more telling, the one-volume summary that was published while the longer version of the dictionary was still in progress included a full page devoted to race.⁶³ Fewer than twenty-five of Kant's terms had more space devoted to them. *Race* was singled out alongside *apperception, concept, knowledge, law, idealism, method, schema,* and *reason.* The word *cosmopolitan*, which received only four pages in the long version, did not even make it into the one-volume edition, though a whole entry was devoted to the *Keime*, drawing on Kant's 1777 essay on race.⁶⁴ Contemporaries wanted to know what Kant meant by *race* because they knew he was using the word in a different sense. In the same vein, it is worth noting that even in the 1820s when Krug (Kant's successor in Königsberg) prepared his dictionary, two-thirds of the entry on "Menschengattung oder Menschengeschlecht" was devoted to Kant's essays on race.⁶⁵ As late as 1833 when Wilhelm Krug published the second edition of his philosophical dictionary "according to the standpoint of contemporary science," he referred extensively to Kant's writings on the subject to explain the technical sense of race.⁶⁶

By introducing the concept of race Kant was not primarily attempting to offer a classification system, although that became a preoccupation of later generations. Nor was the defense of monogenesis, though it was important to him, his only aim. He was looking to offer an account of inheritance. Those who adopted his notion of race without acknowledging the account of *Keime* and *Anlagen* that went with it were left with the same puzzle that led Kant to introduce them in the first place. James Cowles Prichard, the leading theorist of human varieties of the first half of the nineteenth century and an advocate of monogenesis, seems to have addressed the same question that had led Kant to embrace the *Keime* only once, and when he did so, he offered a variation on the same answer provided by Kant. In 1825, in the second edition of his *Researches into the Physical History of Mankind*, Prichard appealed to a theory of germs to explain what was otherwise inexplicable at that time. How could the environment act to establish permanent inheritable characteristics that the environment could not then undo?⁶⁷ Without the *Keime* there was little to distinguish the implications of Kant's account from the implications of Forster's account. The former promoted monogenesis, whereas the latter argued for polygenesis, but what really mattered was that both believed the differences between blacks and whites were now permanent. It should be no surprise that by the mid-nineteenth century polygenesis was dominant, though Kant's theory of germs would never completely disappear.

I have not tried to resolve, once and for all, the complex question of how well Blumenbach, Girtanner, and Schelling understood Kant. I have focused primarily on the task of refuting the claim that Kant must have abandoned the *Keime* once he had read Blumenbach and that this would have led him to abandon the notion of race he had introduced in 1775, either for another different conception of race or to do without race altogether. It is clear that Girtanner and Schelling were not orthodox advocates of Kant's racial theory in all respects. But my point here is only that neither Girtanner nor Schelling hesitated to employ the notion of *Keime* long after (if Sloan was correct) they should have known to abandon it. And even if Blumenbach noticed, who more than anyone—in the conventional reading—should have been offended, so far as I can tell, he issued no public complaint on this issue. I am not denying that Kant's account of race changed after its introduction in 1775, but it seems that we are in danger of exaggerating these changes, if we do not attend to what his contemporaries were saying about it. Blumenbach's *Bildungstrieb* and Kant's *Critique of Judgment* were not the graveyard for Kant's *Keime* that most commentators think they are. Quite the reverse. Until he became aware of Blumenbach's *Bildungstrieb*, Kant had lacked an explanation of how the purposive possibilities inherent in the *Keime* responded to the environment. It was not a perfect match, but there is a danger in thinking that here were irreconcilable differences visible to everyone from the start.

NOTES

1. The importance of this point for understanding Kant's philosophy as a whole will become clearer once currently unpublished work by two colleagues at Pennsylvania State University, Jennifer Mensch, and Mark Fisher, appears in print.
2. For an example of a work that now appears both dated and misleading for precisely this reason, see Helmut Müller-Sievers, *Self-Generation: Biology, Philosophy, and Literature Around 1800* (Stanford: Stanford University Press, 1997).
3. Pauline Kleingeld, "Kant's Second Thoughts on Race," *Philosophical Quarterly* 57/229 (2007):591, n40. I address other aspects of Kleingeld's account in "Kant's Third Thoughts on Race," in *Reading Kant's Geography*, ed. Stuart Elden and Eduardo Mendieta (Albany: State University of New York Press, 2012).
4. I address other aspects of the reception of Kant's writings on natural history among historians of science during this period in, Robert Bernasconi, "The Place of Race in Kant's *Physical Geography* and in the Writings of the 1790s," in *Rethinking Kant*, ed. Pablo Muchnik (Newcastle upon Tyne: Cambridge Scholars Publishing, 2010), vol. 2:268–284.

5. Buffon's belief is found in, Georges-Louis Leclerc, Comte de Buffon, *Histoire naturelle* (Paris: De l'imprimerie royale, 1749), vol. 3:533–534; and, Georges-Louis Leclerc, Comte de Buffon, *Histoire naturelle* (Paris: De l'imprimerie royale, 1766), vol. 14:312–316. For Kant's rejection of Buffon's position and the background to it, see Robert Bernasconi's contribution to a collection of essays on the Kant-Forster debate, "True Colors: Kant's Distinction Between Nature and Artifice in Context," in *Klopffechtereien-Missverständnisse-Widersprüche? Methodische und Methodologische Perspektiven auf die Kant-Forster Kontroverse*, ed. Rainer Godel and Gideon Stienung (Munich: Wilhelm Fink, 2012, pp. 191–207).
6. Immanuel Kant, "Von den verschiedenen Racen der Menschen," *Vorkritische Schriften II, 1757–1777*, Akademie Ausgabe (Berlin: Georg Reimer, 1912), vol. 2:434, trans. Holly Wilson and Günter Zöller, "Of the Different Races of Human Beings," in *Anthropology, History, and Education*, ed. Günter Zöller and Robert B. Louden (Cambridge: Cambridge University Press, 2007), 89.
7. Robert Bernasconi, "Who Invented the Concept of Race? Kant's Role in the Enlightenment Construction of Race," in *Race*, ed. Robert Bernasconi (Oxford: Blackwell, 2001), 11–36.
8. See Staffan Müller-Wille, "Reproducing Difference: Race and Heredity from a *longue dureé* Perspective" in this volume.
9. Immanuel Kant, "Der einzig mögliche Beweisgrund zu einer Demonstration des Daseins Gottes," in *Vorkritische Schriften*, Akademie Ausgabe, vol 2:115, trans. David Walford, "The Only Possible Argument in Support of a Demonstration of the Existence of God," in *Theoretical Philosophy 1775–1770*, ed. David Walford (Cambridge: Cambridge University Press, 1992), 157. Translation modified.
10. Peter McLaughlin, "Kant on Heredity and Adaptation," in *Heredity Produced: At the Crossroads of Biology, Politics, and Culture, 1500–1870*, ed. Staffan Müller-Wille and Hans-Jörg Rheinberger (Cambridge, MA: MIT Press, 2007), 277–291.
11. Kant, "Von den verschiedenen Racen der Menschen," 430; trans., "Of the Different Races of Human Beings," 85. On Kant's racism, see Robert Bernasconi, "Kant as an Unfamiliar Source of Racism," in *Philosophers on Race*, ed. Julie K. Ward and Tommy L. Lott (Oxford: Blackwell, 2002), 145–166.
12. Immanuel Kant, "Bestimmung des Begriffs einer Menschenrace," Akademie Ausgabe, vol. 8:100, trans. Holly Wilson and Günter Zöller, "Determination of the Concept of a Human Race," in *Anthropology, History and Education*, 154.

13. Kant, "Bestimmung," 95; trans., "Determination," 149.
14. Kant, "Von den verschiedenen Racen der Menschen," 430; trans., "Of the Different Races of Human Beings," 85. See also the 1788 account of the concept of race arising as an inference from "a hereditary particularity of different interbreeding animals that does not at all lie in the concept of their species." Kant, "Über den Gebrauch teleologischer Principien in der Philosophie," Akademie Ausgabe, vol. 8:163; trans. Günter Zöller, "On the Use of Teleological Principles in Philosophy," in *Anthropology, History, and Education*, 199.
15. Kant, "Von den verschiedenen Racen der Menschen," 430; trans., "Of the Different Races of Human Beings," 86.
16. Johann Nicholas Tetens, *Philosophische Versuche über die menschliche Natur und ihre Entwicklung* (Leipzig: Weidmanns, 1777), vol. 2:452–476. On Kant and Tetens, see, for example, Arthur Seidel, *Tetens' Einfluss auf die kritische Philosophie Kants, Inaugural–Dissertation zur Erlangung der Doctorwürde der Hohen Philosophischen Fakultät der Universität Leipzig* (Würzburg: Konrad Trilitsch, 1932).
17. Kant, "Von den verschiedenen Racen der Menschen," 434–435; trans., "Of the Different Races of Human Beings," 89–90.
18. Kant, "Von den verschiedenen Racen der Menschen," 435; trans., "Of the Different Races of Human Beings," 90.
19. Indeed, in 1775 Kant employed the phrase "unity of the generative power" (*gemeinschaftlich gültige Zeugungskrafi*) to indicate generative power. Kant, "Von den verschiedenen Racen der Menschen," 429–430; trans., "Of the Different Races of Human Beings," 85.
20. Kant, "Über den Gebrauch teleologischer Principien der Philosophie," 179; trans., "On the Use of Teleological Principles in Philosophy," 214.
21. Kant, "Über den Gebrauch teleologischer Principien der Philosophie," 169; trans., "On the Use of Teleological Principles in Philosophy," 204.
22. Johann Friedrich Blumenbach, *Über den Bildungstrieb* (Göttingen: Johann Christian Dietrich, 1791), 24, trans. A. Crighton as *An Essay on Generation* (London: T. Cadell, 1792), 20. I have cited the text of the 1791 edition, which is virtually identical with that of the 1789 edition. On the 1781 edition see Susanne Lettow, "Generation, Genealogy, and Time" in this volume. For the broader context of the relation between Kant and Blumenbach, see Robert Bernasconi, "Kant and Blumenbach's Polyps: A Neglected Chapter in the History of the Concept of Race," in *The German Invention of Race*, ed. Sara Eigen and Mark Larrimore (Albany: State University of New York Press, 2006), 73–90.
23. Blumenbach, *Über den Bildungstrieb*, 68; trans. *An Essay on Generation*,

57. See also Nicholas Jardine, *The Scenes of Inquiry: On the Reality of Questions in the Sciences* (Oxford: Oxford University Press, 1991), 27.
24. Johann Friedrich Blumenbach, *Über den Bildungstrieb und das Zeugungsgeschäfte* (Göttingen: Johann Christian Dietrich, 1781), 61.
25. For a helpful review of the discussion of the impact of Blumenbach's account of the *Bildungstrieb* on preformationism, see Jörg Jantzen, "Physiologische Theorien," in *Ergänzungsband zu Werke Band 5 bis 9*, Historisch-Kritische Schelling-Ausgabe (Stuttgart: Frommann–Holzboog, 1994), 636–668.
26. Immanuel Kant, *Kritik der Urteilskraft*, Akademie Ausgabe, vol. 5:424, trans. Werner Pluhar, *Critique of Judgment* (Indianapolis: Hackett, 1987), 311.
27. Kant, *Kritik der Urteilskraft*, 423–424; trans. *Critique of Judgment*, 310. Translation modified.
28. Salomon Maimon, *Philosophisches Wörterbuch oder Beleuchtung der wichtigsten Gegenstände der Philosophie in alphabetische Ordnung* (Berlin: J. F. Unger, 1791), 185–189.
29. Norbert Klatt has speculated on why Blumenbach did not publicly attack Kant for his continued use of the term *Keime*. He explains it in terms of their shared position on monogenesis. He also argues that Blumenbach would have been against *Keime*, whether preformed or not, although the evidence is not decisive and needs further research. Klatt also draws attention to the importance of Johann Daniel Metzger's attack on Kant. See Norbert Klatt, "Johann Daniel Metzger und Immanuel Kants Bestimmung des Begriffs der Menschenrasse," in *Kleine Beiträge zur Blumenbach-Forschung* (Göttingen: Norbert Klatt Verlag, 2010), vol. 3:56–62. The importance of Frank Dougherty and now Norbert Klatt's work for revising and deepening our understanding of Blumenbach cannot be underestimated.
30. Although this is not decisive for all of the more technical uses of *Keime*, it should be noted not only that Kant continued to use the notion of *Keime* in a broad anthropological context, but that both he and Blumenbach used it in the context of the development of individuals. In a passage that occurs in all editions of Blumenbach's *Handbuch der Naturgeschichte*, and which is echoed in some student notes for Kant's lectures on physical geography, reason, and language are referred to as *Keime*. Johann Friedrich Blumenbach, *Handbuch der Naturgeschichte* (Göttingen: Johann Christian Dieterich, 1779), 62; 1782, 58; 1788, 59; 1791, 53; 1797, 60; 1799, 61; 1807, 66; 1814, 66; 1821, 66; 1823, 55; 1830, 55. (The 1801

edition was not available to me when I compiled this list.) More significant is Blumenbach's warning that *Keime* is in certain contexts "an indeterminate empty expression," but at this point we need to acknowledge that what is at stake is the difference between Kant's speculative notion of natural history and Blumenbach's conception of natural history, which is, with the exception of the *Bildungstrieb*, largely confined to what Kant would call "natural description." The *Keime* belong to the former. It is Sloan's view that in the 1790s Kant backed away from natural history. I contend that claim in "The Place of Race in Kant's *Physical Geography*," cited earlier.

31. Immanuel Kant, *Kants Vorlesungen*. Band VI, Akademie Ausgabe (Berlin: Walter de Gruyter, 1983), vol. 29:1031; trans Karl Ameriks and Steve Narragun, *Lectures on Metaphysics* (Cambridge: Cambridge University Press, 1997), 497–498.

32. Phillip Sloan, "Preforming the Categories: Eighteenth-Century Generation Theory and the Biological Roots of Kant's A Priori," *Journal of the History of Philosophy* 40/2 (2002):253. For a critique of this essay, see Marcel Quarfood, *Transcendental Idealism and the Organism* (Stockholm: Almqvist and Wiksell, 2004), 102–115. I focus on Sloan because he is probably the preeminent historian of science of this period, and everyone who works in this area is indebted to him.

33. Phillip R. Sloan, "Kant on the History of Nature: The Ambiguous Heritage of the Critical Philosophy for Natural History," *Studies in History and Philosophy of Biological and Biomedical Sciences* 37/4 (2006):627–648. Sloan stipulates that he means by a history of nature "a science that claims to draw warranted conclusions about the developing course of the natural world over time." He explains that this makes it "more than an 'actualist' inference from observed conditions to the workings of a similar set of processes and causes in past time," 629. This differs significantly from Kant's own definition.

34. See Mark Fisher, "Kant's Explanatory Natural History," in *Understanding Purpose. Kant and the Philosophy of Biology*, North American Kant Society Studies in Philosophy, ed. Philippe Huneman (Rochester: University of Rochester Press, 2007), vol. 8:101–121. See also the explanation in John Zammito, "This Inscrutable Principle of an Original Organization," *Studies in History and Philosophy of Science*, Part A, 34, no. 1 (2003). "Kant had to insist that even epigenesis implied preformation: at the origin there had to be some inexplicable (transcendent) endowment, and with it, in his view, some determinate restriction in species variation . . .

This made *epigenesis* into Kant's variant of *preformation*," Zammito, "This Inscrutable Principle," 88. I agree with many of the points made by Zammito in his important and wide-ranging essay. Although I offer a few friendly amendments to his account, the most important function of my essay is to highlight and develop a small part of that story.

35. Kant, *Kritik der Urteilskraft*, 420; trans. *Critique of Judgment*, 306.
36. Kant, "Von den verschiedenen Racen der Menschen," 435; trans., "Of the Different Races of Human Beings," 90.
37. See, for example, John H. Zammito, "Kant's Persistent Ambivalence toward Epigenesis, 1764–90," in *Understanding Purpose: Kant and the Philosophy of Biology*, North American Kant Society Studies in Philosophy, ed. Philippe Huneman (Rochester: University of Rochester Press, 2007), vol. 8:62–64.
38. Friedrich Christian Starke, *Immanuel Kant's vorzügliche kleine Schriften und Aufsätze* (Leipzig: Ernst, 1833), vol. 2:280.
39. For a slightly different perspective from my own, see Robert J. Richards, "Kant and Blumenbach on the *Bildungstrieb*: A Historical Misunderstanding," *Studies in the History and Philosophy of Biology and Biomedical Sciences* 31/1 (2000):11–32. See also Brandon C. Look, "Blumenbach and Kant on Mechanism and Teleology in Nature," in *The Problem of Animal Generation in Early Modern Philosophy*, ed. Justin E. H. Smith (Cambridge: Cambridge University Press, 2006), 355–372.
40. Johann Friedrich Blumenbach, *Handbuch der Naturgeschichte*, 5th ed. (Göttingen: Johann Christian Dieterich, 1797), 23.
41. Christoph Girtanner, *Über das Kantische Prinzip für die Naturgeschichte. Ein Versuch diese Wissenschaft philosophisch zu behandeln* (Göttingen: Vanderhoeck und Ruprecht, 1796), 3–4. I am grateful to Jon Mark Mikkelsen for sharing his partial translation of Girtanner's book with me.
42. Girtanner, *Über das Kantische Prinzip*, 38.
43. Ibid., 54.
44. Immanuel Kant, *Anthropologie in pragmatischer Hinsicht*, Akademie Ausgabe, vol. 7:320; trans. Robert B. Louden, "Anthropology from a Pragmatic Point of View," in *Anthropology, History, and Education*, 415.
45. As Mark Larrimore has pointed out, Kant also continued during the 1790s to republish his essays on race with their references to *Keime*. See his "Antinomies of Race: Diversity and Destiny in Kant," *Patterns of Prejudice* 42, no. 4–5 (2008):358, n30.
46. There has been a tendency in the scholarly literature to play down Girtanner's importance in part because each of his contributions is

isolated from the others. For an overview of Girtanner's writings, see Carl Wegelin, "Dr. med. Christoph Girtanner (1760–1800)," *Gesnerus* 14 (1957):141–163.
47. Girtanner, *Über das Kantische Prinzip*, 54 and 39.
48. Ibid., 25.
49. When it came to skin color, Girtanner drew on Kant's distinction between one's true color, that is to say, one's inherited color, and the artificial color that one had as a result of intervention (such as tattooing)—what we might call lifestyle (particularly pronounced among those who worked outside) and climate. Girtanner translated this distinction into one between the essential, in the sense of the color that is actually proper to a race and that is thus reproduced in the children, and the accidental. To find the true color they should be required to breed, not in the land where they were born and raised, but rather in a foreign country. Girtanner omits the normative role Kant gives to the notion of a temperate climate. Girtanner, *Über das Kantische Prinzip*, 40–41.
50. Johann Friedrich Blumenbach, *De generis humani varietate nativa*, 3rd ed. (Göttingen: Vandenhoeck und Ruprecht, 1795), 68–69; trans. Thomas Bendyshe, *The Anthropological Treatises of Johann Friedrich Blumenbach* (London: Longman, Green, Longman, Roberts, and Green, 1865), 189.
51. Robert Bernasconi, "Kant as an Unfamiliar Source of Racism," 154–158.
52. Sloan, "Preforming the Categories," 251.
53. Kant had written of the "constitution of the skin" rather than its color, but Schelling's change does not alter Kant's sense. Kant, "Bestimmung des Begriffs einer Menschenrace," 106; trans. "Determination of the Concept of a Human Race," 159. Quoted by Friedrich Wilhelm Joseph Schelling in, *Erster Entwurf eines Systems der Naturphilosophie*, in Historische-Kritische Ausgabe (Stuttgart: Frommann-Holzboog, 2001), vol. I, 7:108; trans. Keith R. Peterson, *First Outline of a System of the Philosophy of Nature* (Albany: State University of New York Press, 2004), 44.
54. Kant, "Bestimmung," 105; trans. "Determination," 158.
55. Schelling, *Erster Entwurf*, 109; trans. *First Outline*, 44.
56. Ibid.
57. Ibid., 112; trans. *First Outline*, 48.
58. Schelling, *Erster Entwurf*, 293; trans. *First Outline*, n47. My attempt to show the trajectory that leads from Kant and Blumenbach in Girtanner to Schelling does not mean that I underestimate how this same path helps explain the differences. See Iain Hamilton Grant, *Philosophies of*

Nature after Schelling (London: Continuum, 2006), 97–102. However, in addition to his book on race, Girtanner wrote a number of other works that clearly left their mark on Schelling. Schelling referenced Girtanner's book on chemistry, *Ideen zu einer Philosophie der Natur* (1797), Historisch-Kritische Ausgabe (Stuttgart: Frommann-Holzboog, 1994), vol. I, 5:111–118; trans. Errol E. Harris and Peter Heath, *Ideas for a Philosophy of Nature* (Cambridge: Cambridge University Press, 1988), 59–64. For Girtanner's *Ausführliche Darstellung des Brownischen Systems der praktischen Heikunde* (Göttingen: Johann Georg Rosenbusch, 1798), see Richard Toellner, "Randbedingungen zu Schellings Konzeption der Medizin als Wissenschaft," in *Schelling Seine Bedeutung für eine Philosophie der Natur und der Geschichte*, ed. Ludwig Hasler (Stuttgart-Bad Cannstatt: Frommann-Holzboog, 1981), 117–128.

59. Johann Caspar Lavater, *Essays on Physiognomy* (London: G. G. J. and J. Robinson, [1793]), 103.
60. In addition to the references discussed below, see, for example, *Conversationslexicon*, vol. 4. Cited in Wulf D. Hund, *Rassismus* (Bielefeld: Transcript, 2007), 22.
61. Georg Samuel Albert Mellin, *Encyclopaedisches Wörterbuch der Kritischen Philosophie* (Jena: Frommann, 1801), vol. 4:741–757.
62. Mellin, *Encyclopaedisches Wörterbuch*, 747.
63. Georg Samuel Albert Mellin, *Kunstsprache der Kritischen Philosophie* (Jena: Friedrich Frommann, 1798), 219.
64. Mellin, *Kunstsprache*, 147.
65. Wilhelm Traugott Krug, *Allgemeines Handwörterbuch der philosophischen Wissenschaften* (Leipzig: Brockhaus, 1827), vol. 3:731–734.
66. Wilhelm Traugott Krug, *Allgemeines Handwörterbuch der philosophischen Wissenschaften nebst ihrer Literatur und Geschichte nach dem heutigen Standpunkte der Wissenschaft* (Leipzig: Brockhaus, 1833), vol. 2:844–848.
67. James Cowles Prichard, *Researches into the Physical History of Mankind* (London: John and Arthur Arch, 1825), vol. 2:544. See Hannah Franziska Augstein, *James Cowles Prichard's Anthropology: Remaking the Science of Man in Early Nineteenth Century Britain* (Amsterdam: Rodopi, 1999), 13.

11

SEXUAL POLARITY IN SCHELLING AND HEGEL

ALISON STONE

In this chapter I examine the accounts of sexual difference given by Schelling in his *First Outline of a System of Philosophy of Nature* of 1799 and by Hegel in his mature philosophy of nature, the second volume of his *Encyclopedia of the Philosophical Sciences*, first produced in 1817 and revised in 1827 and 1830. To understand these accounts we must situate them within the broader approaches to nature which Schelling and Hegel adopted in these works. Schelling approaches nature in terms of a fundamental polarity of two forces of productivity and inhibition, a polarity that manifests itself at successive levels throughout the range of natural forms; he considers sexual difference to be the culminating form of this polarity. According to Schelling, the two sexes seek to overcome their polar opposition by reproducing, but they only succeed in generating more finite, sexually differentiated individuals, so that polarity persists in nature indefinitely. Hegel approaches nature in terms of a fundamental opposition between the concept (*der Begriff*) and matter. For Hegel, nature strives to overcome this opposition, doing so increasingly successfully through the range of phenomena from mechanical to chemical to organic. Again, understanding sexual difference in relation to reproduction, Hegel thinks that the sexes reproduce in the effort to realize the (conceptual) unity of their species, but that they only produce another finite, embodied individual. Hegel explicitly aligns the female and male sexes with the material and conceptual sides of the sexual opposition respectively, while Schelling more tacitly aligns the female sex with inhibition and the male sex with productivity.

After reconstructing Schelling's and Hegel's approaches to sexual difference as polarity, I draw some conclusions about the relations between conceptions of sexual difference and the early nineteenth-century project of philosophy of nature (*Naturphilosophie*). First, philosophy of nature as Schelling, Hegel, and others conceived it, made possible a novel conception of sexual difference as the manifestation of fundamental polarities or oppositions within nature. Second, in Hegel's case, his philosophy of nature also allowed him to establish systematic parallels and distinctions between sexual difference and the (supposed) natural racial differences that he considers in his *Philosophy of Mind*, the third volume of his *Encyclopedia*. Third, Schelling's and Hegel's conceptions of sexual difference are ambiguous between the "one-sex" and "two-sex" models identified by the historian of science Thomas Laqueur (1990). Schelling and Hegel treat the sexes as polar opposites, yet in such a way that the female pole is only the negative of the male pole, not a positive term in its own right. As such, philosophy of nature is indirectly implicated in the broader nineteenth-century trend to confine women to the private sphere. Nonetheless, I conclude that feminists can reclaim the philosophy of nature so as to reemphasize the importance and inescapability of sexual difference.[1]

SCHELLING'S POLARIZATION: PRODUCTION AND INHIBITION

In the *First Outline of a System of Philosophy of Nature* (hereafter, the *First Outline*), Schelling situates sexual difference as the culminating manifestation of the polarity of two basic forces (*Kräfte*) structuring all of nature—the productive force and inhibiting force. Moreover, he tacitly understands these two basic forces in sexualized terms, aligning the productive force with the male sex and the inhibiting force with the female sex. To make sense of this, we need some understanding of the project of philosophy of nature that he pursues in the *First Outline*. In turn, this requires tracing how that project took shape in the 1790s, since in the *First Outline* Schelling was reworking (for the third time) the project of philosophy of nature first sketched in his 1797 *Ideas for a Philosophy of Nature*.

Schelling conceived this project in response to Kant's opposition of subjectivity and nature.[2] In the *Critique of Pure Reason* (1781/1787), Kant had argued that ordered experience and knowledge are only possible if the subject of experience applies categories, centrally including that of causality, to the materials of sensation. In turn this application is only possible through the subject's spontaneity or freedom. Yet the categories that make our experience possible constrain us to experience nature—including ourselves insofar as we are part of nature—as a realm of objects whose interactions are causally determined in the manner theorized by Newtonian physics. How then can we exist and act as free subjects within a natural universe in which all events are entirely causally determined? Schelling's solution is that nature already exhibits a form of spontaneity that approximates to the freedom of human agents. "To philosophise about nature," he therefore states, "means to lift it out of dead mechanism . . . to animate it with freedom and to set it into its own free development."[3]

Kant had claimed, however, that ordered experience requires not that we actually *are* free but only that we are capable of thinking of ourselves as free. In his *Critique of Judgment* of 1790, he suggested that this requires us also to think of organisms and of nature as an organized totality, as if they were "purposive"—self-organizing in ways that prefigure the mind's capacity for ordering its own experience. Nevertheless, for Kant this is only a way of thinking about nature. We cannot know whether nature in itself, independently of how we represent it, really is purposive.[4] Thus, for Kant, our judgments of purposiveness are only "regulative," not "constitutive."[5] Schelling, in contrast, believes that we as subjects of experience must really *be* free and that this freedom is only possible if nature really is spontaneous as well—a natural spontaneity that we can therefore know to really exist, independently of our minds. For Schelling, "the purposiveness of natural products dwells *in themselves*, . . . it is *objective* and *real*."[6] He thus aims to describe nature as it objectively, mind-independently is.[7]

In his *Ideas for a Philosophy of Nature*, Schelling further argues that, because we know that in general nature must anticipate human self-determination, we can know about the particular composition of the natural world insofar as we can find its forms and processes to anticipate the self-determining human mind. However, modern scientists have often described these forms and processes in mechanistic terms. It is therefore necessary to reinterpret natural forms and processes to reveal how they prefigure human self-organization. This does not mean that Schelling understands his theory to be an alternative to—or a replacement for—empirical science. On the contrary, he draws throughout on the empirical sciences of his time, reinterpreting their findings in light of his philosophical account of the relation between nature and mind. He also draws particularly on those elements in contemporary science that support his view of nature—while, reciprocally, various scientists including Lorenz Oken and Henrik Steffens begin to undertake empirical research, adopting Schellingian hypotheses as their starting point.[8]

Schelling ascertains from his survey of the sciences of his time that all natural phenomena are best understood as polarized between two forces of attraction and repulsion. On these grounds he criticizes Newtonian atomism, arguing that matter is not fundamental but must be understood as to be composed of prior attractive and repulsive forces:

> Dynamic chemistry . . . admits no *original* matter whatever—no matter, that is, from which everything else would have arisen by composition [as in Newtonian atomism]. On the contrary, since it considers all matter originally as a product of opposed forces (*entgegengesetzter Kräfte*), the greatest possible diversity of matter is still nothing else but a diversity in the relationship of these forces.[9]

These forces, according to Schelling, have the same structure as subjectivity. This is because the subject is inherently oriented simultaneously *both* toward the outer objects in the world about which it tries to know *and* back inward toward itself insofar as whenever we know (or represent) we are, necessarily, implicitly aware that we are doing so. The subject is attracted to know about objects in the world outside it *and* to repel these objects insofar as it returns into self-knowledge.[10] The subject, then, is structured by a polar opposition between forces of attraction (or expansion) and repulsion (contraction). Since contemporary science shows—or can be reinterpreted as showing—that nature is structured by a basic opposition between attraction and repulsion, scientific research confirms that nature is structured in the same way as the subject, prefiguring its freedom.

We see, then, how Schelling came to conceive of nature as fundamentally organized by two forces: attraction, which is outward-oriented and expansive, and repulsion, which is inward-oriented and withdrawing. These forces have tacit sexual connotations, which become relatively explicit when Schelling subsequently reconceives the forces in terms of production and inhibition in the *First Outline*.

Male Production — Female Inhibition

After advancing a second version of *Naturphilosophie* in *On the World-Soul* (1798), in which Schelling argues that the basic forces composing nature can only exist if nature is purposive as a whole and in that sense has a "soul,"[11] he quickly moved on to the 1799 *First Outline*. Its premise is that nature is originally productive. Nature originally consists in sheer, unlimited, productive activity (*unendliche productive Tätigkeit*). This productivity "limits" (or "fixates," *fixirt*) itself to constitute the various particular products and processes that make up the natural world. These products, Schelling insists, are not permanently fixed entities but are only transitory resting points within nature's unending productivity—like eddies in a stream.[12] According to Schelling, the mistake of much mechanistic empirical science is to overlook the underlying productivity that first makes these products possible, "the inner driving activity [*Triebwerk*]."[13] His aim in the *First Outline*, then, is to analyze nature's free productivity and trace how it develops, and in this light, to reinterpret the various finite natural products studied by the sciences.

The key question that Schelling confronts is *how* nature's productivity becomes confined in particular products. Infinitely active as it is, nature's productivity would pass through an endless array of products infinitely quickly, destroying each product as quickly as it had been created, unless that productivity encountered some "retarding" force.[14] The distinction between productivity and products must be explained by a prior *Dualität* of underlying forces in nature—a duality of productivity and another force opposing it. To use Schelling's analogy, a river only forms eddies when its flow encounters resistance.[15] He infers that a force of inhibition (*Hemmung*) must oppose nature's productivity so that particular products arise through the resulting conflict of productive and inhibiting forces. Each product or process reflects a particular level of equilibrium between these opposed forces: "Each formation is itself only the . . . appearance of a determinate proportion which nature achieves between opposed, mutually limiting actions."[16] Each natural product is structured by a polarity: it reflects at once a given level of productivity and a given level of inhibition. We recognize these two forces as the latest version of the forces of attraction and repulsion—attraction that expands outward in ever-new activities and repulsion that pulls the first force back into the determinate shape of a product.

Schelling proceeds to reinterpret scientific accounts of physiology, magnetism, electricity, and chemistry, to see how processes in each of these domains manifest the polarity of forces. In his view, this array of manifestations of polarity arises because natural productivity bursts beyond each polar product in which it becomes confined. Being infinitely productive, it must always transcend its limitations.[17] But then productivity must become inhibited again so that a whole series of products results. For Schelling, sexual difference, *Geschlechtsverschiedenheit*, is at the apex of this series. It is a difference found, beyond humanity, throughout the entire organic realm of animals and plants, albeit in different forms and not always distributed across different individuals: "Throughout the whole of [organic] nature absolute sexlessness is nowhere demonstrable."[18]

This must be because sexual difference is necessary for reproduction, Schelling infers. In turn, he can make sense of this necessity in terms of his basic conception of nature. Difference, polarization into two forces, makes possible nature's productivity as a whole—nature can be creative, generative, only on condition of being divided into two polar forces. Likewise, then, organic individuals must be able to regenerate only on the condition of succumbing to the same polar division. As Schelling puts it, "the separation into different sexes is just the separation which we have furnished as the ground of inhibition in the productions of Nature."[19]

The division of organic beings into two sexes, then, arises so they can reproduce. But why do organic beings seek to reproduce? Presumably their urge to do so manifests the productive force within them, which drives them to try to pass beyond their finite boundaries in a creative way.[20] Insofar as organic beings seek to reproduce so as to overcome their finite boundaries, then, they must also seek by reproducing to overcome the division into two sexes, within which they have become confined. It is in this respect that nature "hates" sex, Schelling writes.[21] This is also why Schelling maintains that living beings seek by reproducing to realize the unity of their entire species—their *Gattung*, to be realized in reproduction as *Begattung*.[22] The productive force that these finite individuals seek to realize takes the form, at this point in nature, of the species as a transindividual unity into which these individuals endeavor to submerge their differences.

However, Schelling maintains, reproductive activity never succeeds. It only ever issues in new finite products: the couple's offspring—finite both in body and in being sexed. This failure reflects the fact that natural productivity can never get free of inhibitive force. Whenever productivity tries to release itself, it necessarily becomes inhibited again. The offspring, then, must remain finite and sexed and they cannot embody the resolution of the opposition of natural forces; otherwise, their appearance would put an end to all the striving and activity in nature.[23]

Schelling has given sexual difference central importance within nature as the culminating manifestation of nature's basic polarity. He does not, however, elaborate on the nature of the two opposed sexes. Yet his broader contrast between the forces of productivity and inhibition inescapably takes on tacit gendered connotations, given the entire history of gendered philosophical contrasts against the background from which Schelling writes.

Productivity—which he also calls nature's "subjectivity"[24]—is symbolically male in virtue of its connections with activity, mind, and power; inhibition is symbolically female in virtue of its connections with withdrawal, passivity, and interiority. These contrasts become relatively explicit at one point in his *First Outline* when Schelling interprets sexual difference in terms of the difference between "receptivity" and "irritability." He is referring to John Brown's medical theory, according to which disease arises from a mismatch between an individual's inherent level of "irritability" and the level of stimulation impinging on them from the environment. Schelling interprets irritability as a form of productive force and receptivity as a form of inhibiting force. In his view, children are highly susceptible to stimulation—highly receptive—but are correspondingly lacking in irritability; that is, in terms of his broader account of nature, in children the inhibiting force predominates over the productive force. He adds here:

> If the organic power of resistance increases, the movements become more forceful, more energetic too—in equal proportion to the sinking sensibility.—Or, one might observe the difference of the sexes, or the climatic differences of peoples, or finally the increase of the forces directed outwardly in nature, which also happen in a certain (inverse) relation to sensibility.[25]

Sexual difference, then, is taken to exemplify how sensibility (inhibition) may predominate over irritability (productivity) and, equally, how irritability may reassert itself against sensibility. Schelling does not say which sex embodies which possibility, but we can assume that for him the female sex embodies a predominance of sensibility and the male embodies the reassertion of irritability.

We can also see that Schelling's productive and inhibiting forces have this gendered significance by recalling the enormous influence of his 1794 readings of Plato's cosmological dialogue the *Timaeus* on his thought.[26] For Plato, the construction of the cosmos depends on the existence of the formless, primal, material space that he calls *chora* and explicitly describes in feminine terms, as the receptacle and nurse of generation. The *chora* corresponds to the inhibiting, withdrawing, contracting force that Schelling later identifies within nature—a force that carries over the *chora*'s feminine connotations. Indeed, when he subsequently reformulates the idea of the withdrawing force in his 1809 essay on human freedom, Schelling equates it with the "maternal body . . . the obscurity of that which is without understanding . . . the mother of knowledge."[27]

These gendered contrasts imply that the male function in reproduction is actively to initiate sex and reproduction and to embody the striving of natural productivity beyond the fixed forms into which it has become confined. In contrast, the female function is to subject this expansive, productive force to renewed inhibition, presumably by confining the unity of the species within the finite form of embodied individual offspring. The male creates the active, formative principle of the offspring, while the female encloses and puts flesh on this male creation. These ideas remain only implicit in Schelling; however, Hegel theorized the respective natures of the male and female sexes more systematically.

HEGEL'S POLARIZATION: CONCEPT AND MATTER

Hegel discusses reproduction—the "species-process" (*Gattungsprozess*)—in §368–§369 of his philosophy of nature, the second part of his *Encyclopedia*. However, in these two main paragraphs he does not refer to sexual difference, which he considers only in the "addition" to §368. Hegel's editor Jules Michelet assembled this, as with all the additions to the philosophy of nature, from various sources including student transcripts, Hegel's Heidelberg and Berlin lecture notes on nature spanning 1819 to 1830, and his Jena lecture notes on nature and mind dating back to 1805 to 1806. It is from the Heidelberg, Berlin and, above all, the Jena notes[28] that Michelet drew Hegel's account of sexual difference.

It might seem that we cannot wisely interpret Hegel's mature conception of sexual difference based on passages largely composed of material dating back to 1805 to 1806. After all, he did not see fit to include an account of sexual difference in the main paragraphs of his mature philosophy of nature, and he did not in his maturity give sexual difference the same prominence as he did at the time of his Jena drafts (for instance, sexual difference is only mentioned briefly in the transcript of Hegel's 1823/24 nature lectures made by K. G. J. von Griesheim). However, this does not necessarily mean that Hegel had ceased to be concerned about the nature of sexual difference or to uphold broadly the same account of it that he first sketched in Jena. It is notable that in his 1821 *Philosophy of Right*, Hegel appeals to his philosophy of nature to support his claims about the proper social roles of men and women. Infamously maintaining that women's place is in the home—"Woman [*die Frau*] . . . has her substantial vocation in the family, and her ethical disposition consists in this *piety*"[29]—he argues that it is through this gendered division of social roles that, "The *natural* determinacy of the two sexes acquires an *intellectual* and *ethical* significance."[30] What is this "natural determinacy" of the sexes? Hegel tells us that it arises out of "life in its totality, . . . as the actuality of the *species* and its process,"[31] thereby referring us to his discussion of the species-process in §368–§369 of his philosophy of nature. Clearly, Hegel understood his account of reproduction in those paragraphs to identify and explain the distinct natures of the two sexes and so to provide a basis for his sociopolitical division of gender roles. This suggests that Hegel essentially retained the account of sexual difference that he first worked out in Jena—an account that ties in with his treatment of women in the *Philosophy of Right*. The addition to §368 may thus be treated as presenting Hegel's considered and ongoing understanding of sexual difference.

The context of this view is Hegel's account of "sexual relationships" (*Geschlechtsverhältnisse*)—by which he means the reproductive activities of animals, including human beings considered solely in respect of the characteristics they share with animals. Sexual relationships arise when one animal encounters another of the same species. These encounters are the first case that Hegel finds within nature, in which one subject enters into relationship with another. Each animal senses that the two are both "identical"—insofar as they belong to the same species—and "different"—as individuals. Each animal senses a tension between the identity and the difference: it "has the *feeling* of this defect [or tension]. The species [*Gattung*] is therefore present in the individual as a straining against the inadequacy of its single actuality."[32] Each animal therefore acquires an urge to realize the identity of the two by copulating with the other and producing offspring in which this identity will be embodied. "In the natural state the identity of the sexes is . . . a third, that is *produced*, in which both sexes intuit their identity as a natural actuality."[33]

As we can see, Hegel has retained Schelling's view that reproduction strives to realize the species as a unity opposed to the individuality of the reproducing animals. Ultimately, though, Hegel argues—again like Schelling—that reproduction always fails. The offspring are still individual animals, who differ from their parents as yet more individuals, and who become compelled to pass through the same reproductive process, as will their own offspring, and so on *ad infinitum*.

To see how Hegel's account of sexual difference derives from this theory of reproduction, we must spell out certain assumptions that Hegel makes here—assumptions that he does not make explicit, but which we may impute to Hegel insofar as, by doing so, we can make intelligible his approach to sexual difference. Hegel assumes that in any reproductive process the two participant animals must play different roles. Reproduction is a process with a *telos* or purpose, which produces a third entity that incarnates the identity of the two animals that contribute to it. Just as each organism *qua* purposive whole must articulate itself into specialized subsystems,[34] likewise, the two individuals carrying out the purposive activity of reproduction must assume specialized roles within it. The entity to be produced must be a "third," different from the parents, so one parent must have the role of producing the child as a distinct individual. Yet the offspring is also to be nothing more than an embodiment of the identity between the parents. In this respect, the offspring must be identical with the parent(s). It falls to the second parent to produce the offspring as something identical with the parent(s).

Each parent animal develops a specific reproductive anatomy that enables it to play one or the other of these roles. "The *formation* [that is, the anatomical shape] *of the differentiated sexes* must be different, their determinacy against each other which is posited by the concept [that is, which is logically required] must exist."[35] Regarding male (*männliche*) animals, Hegel states that by lying on the body's exterior, their genitals embody "the moment of duality [*das Entzweite*], of opposition."[36] It is distinctive of male genitals that they are primarily located on the outside of the body. Generally, Hegel believes that "external" organs and limbs enable animals to engage and interact with items in the external world. He holds that the outward development of an animal's anatomical shape reflects its "connection with an other outside it."[37] The "other" to which male animals are related in the reproductive process is the species as-it-is-to-be embodied in the offspring. Thus, male genitals have the form they do because these genitals enable the animal to play the role of relating to its offspring as to something that is other to (or different from) it. This anatomy enables male animals to contribute to the offspring in a way that treats it as something different from the male parent—by expelling it outside that parent's body, in the shape of semen.

For Hegel, those animals whose role is to produce the offspring as something identical with them develop a female (*weibliche*) anatomy. Characteristic of the female genitals is that they are located on the inside of the body. "The male testicle remains enclosed in the ovary in the female, does not emerge into opposition."[38] Their internal anatomy allows females to contribute to their offspring in a way that treats the offspring as something identical to them—a part of their own bodies. Their anatomy allows females to retain their offspring in their wombs, within their own bodies, as part of their own bodily processes. Thus, "the female remains in her undeveloped unity."

Hegel sums up: "in one or other of these genitals, one or the other part is essential; in the female this is necessarily the undifferentiated element [*das*

Indifferente], in the male, the moment of . . . opposition."[39] Hegel's idea is that female bodies are organized by a principle of self/other unity. Female anatomy reflects and realizes a reproductive role in which the mother and her offspring form an undifferentiated unity, with no firm boundary between the mother's body and that of her offspring. The female body embodies immediate unity between self and other; the male body, in contrast, embodies difference between self and other.

Whereas Schelling tacitly associated inhibition with the female and productivity with the male, Hegel instead associates the female with lack of difference and the male with difference. Nonetheless, like Schelling, he regards this sexual opposition as one of the highest-level manifestations of an opposition that has structured the entirety of nature. As we would expect, for Hegel this is no longer an opposition of productive and inhibiting forces. Indeed, Hegel abandons the terminology of forces and instead understands nature to be organized by two dimensions or aspects—the concept and matter. To see this, we need to briefly survey the whole course of his philosophy of nature.

The Sexualization of Concept and Matter

Hegel begins the text by presupposing that nature has emerged from the "idea," itself understood as the highest form of the "concept"—which is the whole rationally interconnected sequence of basic ontological principles and forms (being, nothingness, becoming, determinacy, etc.), the development of which is narrated in Hegel's *Logic*, the first volume of his *Encyclopedia*. The Hegelian concept, then, is nothing subjective but is, rather, "the true, objective, actual being of things themselves. It is like the Platonic Ideas, which . . . exist in individual things as their substantial kinds."[40] At the end of Hegel's *Logic*, the idea comes out of itself, or externalizes itself, to constitute nature. As the product of the idea's self-externalization, nature initially exists as sheer "externality," *partes extra partes*—that is, pure matter. Because the idea becomes absolutely other to itself, assuming a character utterly other to its inherent character of pure rationality and articulated unity, the idea enters into the shape of matter, that which is utterly nonrational and is a pure manifold of mutually indifferent elements.

Hegel then traces how the matter, of which nature initially consists, becomes permeated by the concept in a "series of stages consisting of many moments, the exposition of which constitutes the philosophy of nature."[41] As with Schelling, Hegel is not putting forward a theory to replace or compete with empirical science. Rather, Hegel constantly draws on and reinterprets empirical scientific hypotheses and observations, recasting and reorganizing them in light of his guiding philosophical idea of the opposition between concept and matter. First, he maintains that the concept reemerges within nature in the form of unifying principles that hold portions of matter together into individual bodies. Second, the concept increasingly reshapes matter into forms that express and reflect it, so that material bodies acquire increasingly complex properties—first mechanical, then electrical, then chemical—in respect of which their matter progressively comes to manifest the complex, articulated character of the concept. At the pinnacle of nature's hierarchy stand animals, whose bodies are completely conceptually permeated: "the whole [of the animal's body] is so pervaded by its unity that . . . in the animal body the complete untruth of [material] being-outside-one-another is revealed."[42]

Now, all along, Hegel understands matter to be symbolically female, a symbolic equation that surfaces explicitly in several places. One is the following passage from the "Introduction" to the *Philosophy of Nature*:

> The study of nature is . . . the liberation of spirit in nature . . . This is also the liberation of nature . . . Spirit has the certainty which Adam had when he looked on Eve, "This is flesh of my flesh, and bone of my bone." Thus Nature is the bride which spirit weds.[43]

By tracing how the material side of nature becomes increasingly permeated by its conceptual side, the Hegelian philosopher "liberates" nature and establishes that it is of the same "flesh" as the philosopher: nature is not pure matter standing over against us as beings of pure mind; rather, we are composed of concept-permeated matter just as nature is. This places nature in the same relation to human beings as Eve to Adam in the book of Genesis: Eve and nature share in the concept-permeated materiality of Adam and humankind. Nonetheless, nature remains *relatively* material compared to humanity—for, in much of nature, the concept struggles to express itself within matter. By implication, Eve, too, is relatively material compared to Adam. As a female, Eve is more material than Adam and, implicitly, her greater materiality is what *makes* her female and not merely another man. Matter is symbolically female, so that those individuals who are more material are thereby qualified as female.

But in what sense can some individuals be more material than others? An answer is provided by Hegel's account of reproduction and sexual difference.

Having described male and female reproductive anatomies to be organized respectively around difference and its absence, he adds that female and male individuals respectively contribute "the material element" and the "subjectivity" to their offspring.

> *Procreation* must not be reduced to the ovary and the male semen, as if the new product were merely a composition of the forms or parts of both sides; the truth is that the female contains the material element, but the male contains the subjectivity.[44]

Because the female retains the fetus within her own body, as part of her own flesh, she exchanges bodily materials with the fetus on an ongoing basis, in that respect contributing to the fetus materially. In contrast, the male expels semen; having done so, he has no further material, corporeal relationship with the fetus. How is the male thereby bestowing subjectivity on the fetus? Hegel explains: "The seed is ... [a] simple representation ... —simply a *single* point, like the name and the entire self."[45] Because the male's contribution to the fetus takes the form of one single emission of semen (rather than many material exchanges over time), the male is providing matter in a shape suited to represent the individuality of the fetus: the single emission of semen "represents" the child-to-be as a single individual. As such, the material shape of the emission of semen reflects the concept: the matter that the male provides towards the fetus is concept-permeated. Hegel can therefore extrapolate that the male bestows upon the fetus subjectivity or mind in latent form. Meanwhile, the female contributes matter that does not reflect the same principle of subjective unity but remains dispersed in a multiplicity. She contributes matter in a form that is less reflective of (and permeated by) the concept. In her contribution to the fetus, matter predominates; in the male contribution, the concept predominates.

Hegel tacitly equated matter with the female and the concept with the male throughout his account of nature. It is unsurprising, then, that when he theorizes sexual difference, he maps male/female difference onto that of concept and matter. Ultimately, for Hegel, sexual difference manifests the opposition of concept and matter that has organized all nature. Whereas for Schelling, sexual difference manifested the opposition of two basic forces, for Hegel it manifests the opposition of nature's two organizing dimensions. For both thinkers, though, sexual difference reflects nature's fundamental constitutive opposition.

Sexual and Racial Difference

Hegel's reconception of sexual difference in terms of the dynamic conflict between concept and matter also allowed him to place sexual and racial difference systematically in relation to one another. He discusses racial difference within his *Philosophy of Mind*, the third part of his *Encyclopedia*, under the heading of "Physical Qualities" of the "Physical Soul," as Wallace translates it—or, more accurately, "Natural Qualities," *natürliche Qualitäten*, of the "Natural Soul," *natürliche Seele*.[46] The soul is the first form of spirit in Hegel's architectonic. Advancing beyond animal life, in the (human) soul the unity of the concept returns to itself within its material body, whereby we have the beginnings of mind.[47] Yet at first the content of this conceptual unity is completely given by the material body out of which the conceptual unity returns to itself as mind. The mind does not yet reshape this matter after its own model, as it will come to do at more advanced stages by thinking, speaking, generating moral and intellectual principles, reasoning, and so on. As yet, the mind as soul remains immersed in its material body and surroundings, namely the physical earth and environment in which each body is located. Here Hegel infers that:

> The universal planetary life of the nature-governed mind [*Naturgeistes*] particularises itself according to the concrete differences of the earth and breaks up into the particular nature-governed minds [*besonderen Naturgeister*], which give expression to the nature of the geographical continents and constitute *racial diversity* [*Rassenverschiedenheit*].[48]

The difference between the human races is still a natural difference, that is, a difference that immediately concerns the natural soul. As such, the difference is connected with the geographical differences of those parts of the world where human beings are gathered together in masses. These different parts are what we call continents.[49]

Hegel proceeds to identify three principal races: the Negroes of the African continent, allegedly marked by childish *naïveté*—in effect, they remain in immediate unity with nature; the Mongols of the Asian continent, who rise to form a concept of spirit, that is, of the divine that lies beyond nature, but who remain limited by treating the divine as exclusively beyond and opposed to nature; and the Caucasians of the European continent, who progress to recognize that divine spirit also exhibits itself within nature, and who thereby attain to a concept of spirit as self-determining—in turn becoming able to exercise practical self-determination and so make world history, as the other races cannot (not at all in the case of Negroes, and only very imperfectly in the case of Mongols).[50]

Deeply problematic as this explicitly hierarchical classification of the races is, here, I merely want to draw out the set of systematic connections and distinctions that Hegel establishes between racial and sexual difference. For him, as we have seen, sexual difference arises just where nature attempts to overcome its constitutive opposition (as Schelling likewise thought). As such, for Hegel as for Schelling, sexual difference is a dynamic opposition, one that is constituted by the attempt to overcome itself, even though it fails in this attempt. In contrast, for Hegel, racial difference is a difference to which the human soul becomes subject insofar as it remains immersed within its natural surroundings in an immediate way. Racial difference is not a dynamic, self-transcending opposition but a form of diversity to which the concept (in its highest form, the human mind) succumbs insofar as it has *failed* in its attempt to transcend its opposition to matter by remodeling matter after its own image. For Hegel, nature in general is a realm of material multiplicity—a realm in which the concept does not fully predominate over matter but is endlessly dispersed into the manifold materials in which it only imperfectly manifests itself. The same is true of the material environment of the earth: because it is natural and material, it contains a manifold of differences that have not been subordinated to the articulated unity of the concept. These differences in turn infect the human soul, rendering it racially differentiated.

In fact, there is a systematic parallel between racial difference and the division of animal kinds (*Gattungen*) into manifold species (*Arten*), with which Hegel deals in §370 of the philosophy of nature. Having outlined how reproduction fails to overcome differences between the sexes and between animals as embodied individuals, Hegel now explains that, because animals inevitably remain embodied, they can never embody their kinds perfectly. Instead animals become subject to multiple variations, owing to "the manifold conditions and circumstances of external nature."[51] Indeed, these differences infect animal kinds themselves so that these become subdivided into species. At the conceptual level, the division of species results from the concept's failure to master matter in reproduction.

Differences between species parallel the racial differences that, for Hegel, arise later in the systematic development of the concept and mind. Whereas sexual difference manifests the unsuccessful effort of the concept to overcome its opposition to matter through reproduction, species difference manifests the reality that this opposition cannot be overcome within nature, so that the concept always remains subject to material diversity. Likewise, racial difference reflects the fact that the opposition of concept (now existing as mind) and matter still cannot be overcome at the level of the natural soul, so that mind as natural soul remains subject to the material diversity that prevailed within nature.

SEXUAL DIFFERENCE AND THE PHILOSOPHY OF NATURE

Peter Hanns Reill has argued that the *Naturphilosophen* generally gave sexual difference an important place within their theories of nature because they took nature to be organized by polar oppositions, and because sexual difference could readily be interpreted in terms of these kinds of oppositions, so that it could be construed as epitomizing and confirming the pervasive operation of natural polarity.[52] We have seen this with regard to both Schelling and Hegel. Likewise, Philippe Huneman has recently observed that a "*leitmotif* of the hermeneutics of nature is that of gender (*Geschlecht*) as a major sign of nature's finitude—as nature's attempt and failure to overcome such a finitude."[53] Gender, or sexual difference, arises where nature attempts and fails to overcome its constitutive polar opposition by way of reproduction.

The interpretation of nature as organized by polar oppositions, then, meant that sexual difference assumed new importance for the philosophers of nature. Its existence appeared to provide one source of empirical confirmation of their view that polar oppositions pervade nature. Moreover, their view of nature led the *Naturphilosophen* not only to highlight the importance of sexual difference but also to understand it in new ways, in terms of the various kinds of polar opposition that each thinker took to underlie nature—between productivity and inhibition for Schelling, concept and matter for Hegel. Sexual difference became understood *as* an opposition[54] reflecting and condensing the dynamic and conflictual structure of the whole natural world.

By interpreting sexual difference as a polar opposition, Schelling and Hegel reconceived it in a way that was peculiarly appropriate to their time. This was a time when, as Thomas Laqueur has shown, the "one-sex" model of sexual difference that had prevailed in the West ever since the classical period was becoming supplanted by a new "two-sex" model, a biological model on which the sexes were radically and completely different. This conceptual transformation began in the late eighteenth century, as Laqueur and Londa Schiebinger (1989) show. On the "one-sex" model, female genitals are essentially the same as the male's except that they are on the inside of the body. The vagina counts as an internal penis; the uterus, an internal scrotum. According to the Aristotelian biology that had prevailed for centuries, women have these internal male organs because their bodies have less heat with which to expel their organs outward. In the late eighteenth century, scientists began to reconceive the female body as radically different from the male body, every element of female anatomy manifesting this radical difference. Female anatomical structures were given their own names, and the first anatomical drawings of the female skeleton in its sexual specificity were made.

These conceptual changes reflected the *political* upheavals of the day. In the wake of the epoch-making event of the French Revolution, long-standing assumptions about divinely ordained metaphysical and social hierarchies could no longer be taken for granted. On those old assumptions, women's metaphysical and social status as men's inferiors is reflected anatomically in that the female body is merely an inferior, inadequately developed and exteriorized version of the male body. Because the French Revolution had discredited these old ways of thinking, Schiebinger argues, new justifications for patriarchy were sought. It is within this context that scientists—then almost entirely male—sought to show that the radical difference of the female body disqualified women from public life. Female skeletons were depicted with large pelvises suiting them for child-bearing and family life.

As we might expect from their historical location, Schelling and Hegel understand the sexes in ways ambiguous between the one- and two-sex models. To be sure, it might seem that Schelling regards the sexes as radically different, insofar as he maps male and female onto productivity and inhibition as two basic forces, each irreducible to the other. However, in the *First Outline* Schelling equivocates on whether these forces are equally basic, as David Krell (2002) shows. Sometimes Schelling claims that there is an *original* diremption (*Entzweiung*) into two forces within nature[55]—an *original opposition.*[56] At other times, he regards inhibition as secondary—merely the form taken by productivity when it turns against and restricts itself as it must, if it is not to dissipate and squander its own creative activity.[57] At these points Schelling sees inhibition merely as the negative, inverted form of original productivity—just as, on the one-sex model, the female sex was regarded as merely the inverted negative of the male. Ultimately, then, Schelling is ambiguous about whether female inhibition is radically different or whether it is merely the negative form of male productivity.

Hegel appears to subscribe more clearly to a one-sex model, stating that female organs are merely inner versions of male organs.[58] For him, both sexed anatomies are organized by the shared purpose of realizing the genus in material shape, but male anatomy is oriented toward difference, hence outward; while female anatomy is oriented toward unity, hence inward. However, as we see here, Hegel draws these one-sex conclusions from a view much closer to the two-sex model of radical difference, on which female anatomy is entirely organized by a principle of unity with the species, whereas male anatomy is organized by individual difference from the species. On this view, every detail of the sexes' respective anatomies reflects this fundamental difference in their purposive natures. After all, then, Hegel thinks—as on the two-sex model—that a radical difference in organizing principles manifests itself in every facet of sexed anatomy and embodiment. Like Schelling, he equivocates between the one- and two-sex models.

Either way, the conceptions of sexual difference made available by German idealist philosophy of nature might appear to hold little appeal for contemporary feminists. These conceptions combine features of the one-sex and two-sex models, both of which, in different ways, justify women's social subordination to men. Hegel's conception is also intricately connected with his racial hierarchy. Even so, I believe that there are ways in which it is worthwhile for feminists today to think with the philosophy of nature and its conceptions of sexual difference. In particular, as we have seen, sexual difference assumed central importance within the philosophy of nature. The idea that polar forces, or basic oppositions, organize nature meant that sexual difference could be seen as nature's culminating manifestation. However, for some critics such as Reill,[59] this makes *Naturphilosophie* complicit with the nineteenth-century insistence on sexual opposition as a justification for confining women to the private sphere. Reill refers especially to the accounts of sexual opposition given by the late *Naturphilosoph* Carl Gustav Carus and the follower of Schelling Lorenz Oken. Oken effectively materialized Schelling's polarization of productivity and inhibition, translating these forces into empirical entities: the productive male polyp that impregnates the receptive female plant. Reill further suggests that *Naturphilosophie* reacted against a late eighteenth-century move to valorize androgyny and sexual mixing, an intellectual move that corresponded to the rise of relatively relaxed and sexually mixed modes of social intercourse, which enabled notable women of the time such as Rahel Varnhagen, Caroline Schlegel-Schelling, and Dorothea Veit to rise to prominence.

Yet in fact these ideals of androgyny and mixing were not unequivocally favorable to women. These ideals relied on a conceptual contrast between the male and female principles that the most fully developed human character is supposed to combine and mix—and according to this contrast, the male generally is active, productive, or intellectual whereas the female is passive, receptive, or emotive.[60] Although according to these ideals empirical women can and should incorporate male traits, the underpinning symbolic contrast between male and female elements still places the female strands in negative opposition to the male. But since women *qua women* are necessarily linked to what is symbolically female, women remain in an inferior position in this scheme—culturally and, in consequence, socially. In this respect the late-eighteenth-century notion of androgyny is not so sharply differentiated from the *naturphilosophisch* emphasis on sexual polarity as Reill suggests.

In any case, an emphasis on sexual difference need not be reactionary but can support the work of contemporary feminist theorists who likewise stress the importance of sexual difference. As Elizabeth Grosz (1990) among others has insisted, we do not think, reason, or act as sex-neutral beings, much as we may try to do so. We are sexually differentiated beings, and the attempt to escape or transcend this reality in thought or practice can only ever be deluded. This feminist insight, as articulated by Grosz and others, has generated the important project of tracing how sexual difference has left its mark on our intellectual and cultural productions—and how, in particular, these have often been marked by the male sex of their creators. An equally important feminist project is that of creating new bodies of thought, knowledge, and practice out of women's sexually specific forms of embodiment. Because the philosophy of nature makes sexual difference central within nature, it can provide a justification for this feminist insistence on the constitutive force of sexual difference. It can also provide us with fresh perspectives on how our sexually differentiated forms of embodiment inescapably shape our thinking and activities, insofar as the mind is always a mind within nature.

NOTES

1. In arguing that the philosophy of nature remains worthwhile for feminists, I differ from Peter Hanns Reill, who unfavorably contrasts the *naturphilosophisch* emphasis on gender polarity to (what he sees as) the late-Enlightenment emphasis on gender mixing and androgyny. I address Reill's contrast below.
2. On the relation of Schelling's project to that of Kant see, among others, Frederick Beiser, *German Idealism: The Struggle Against Subjectivism 1781–1801* (Cambridge, MA: Harvard University Press, 2002), part IV; Andrew Bowie, *Schelling and Modern European Philosophy* (London,

New York: Routledge, 1993); George di Giovanni, "Kant's Metaphysics of Nature and Schelling's *Ideas for a Philosophy of Nature*," *Journal of the History of Philosophy* 17 (1979):197–215; Robert J. Richards, *The Romantic Conception of Life: Science and Philosophy in the Age of Goethe* (Chicago: University of Chicago Press, 2002), esp. 128–145.

3. Friedrich Wilhelm Joseph Schelling, *First Outline of a System of Philosophy of Nature* (1799), trans. Keith R. Petersen (Albany: State University of New York Press, 2004); German version: *Erster Entwurf eines Systems der Naturphilosophie*, in *Werke: Historisch-kritische Ausgabe*, ed. Wilhelm G. Jacobs and Paul Ziche (Stuttgart: Frommann-Holzboog, 2001), vol. I, 7:14/79. References to repeatedly cited works by Schelling and Hegel (in English translation) are parenthetical and are sometimes amended following the German texts without special notice. Paragraph number precedes English pagination.

4. Immanuel Kant, *Critique of Judgment* (1790), trans. Werner S. Pluhar (Indianapolis: Hackett, 1987), §65–§67, pp. 255–259.

5. Kant, *Critique*, §67, p. 259.

6. Friedrich Wilhelm Joseph Schelling, *Ideas for a Philosophy of Nature* (1797), trans. Errol E. Harris and Peter Heath (Cambridge: Cambridge University Press, 1988); German version: *Ideen zu einer Philosophie der Natur*, ed. Manfred Durner, *Werke: Historisch-kritische Ausgabe* (Stuttgart: Frommann-Holzboog, 1994), vol. I, 5:32/96.

7. Bowie, *Schelling*, 31.

8. See, inter alia, Richards, *Romantic Conception*.

9. Schelling, *Ideas*, 221/252.

10. Ibid., 176/214–215.

11. See Richards, *Romantic Conception*, 139.

12. Schelling, *First Outline*, 32/98.

13. Friedrich Wilhelm Joseph Schelling, "Introduction to the Outline of a System of the Philosophy of Nature, or, On the Concept of a Speculative Physics and the Internal Organization of a System of this Science," in *First Outline of a System of Philosophy of Nature* (1799), trans. Keith R. Petersen (Albany: State University of New York Press, 2004); German version: "Einleitung zu dem Entwurf eines Systems der Naturphilosophie," in *Schriften (1799–1800), Werke: Historisch-kritische Ausgabe*, ed. Manfred Durner and Wilhelm G. Jacobs (Stuttgart: Frommann-Holzboog, 2004), vol. I, 8:196/32.

14. Schelling, *First Outline*, 187/266. See also Bowie, *Schelling*, 36.

15. Schelling, *Introduction*, 206/45–46.
16. Schelling, *First Outline*, 35/101.
17. Ibid., 140/209.
18. Ibid., 36/102.
19. Ibid., 39/105.
20. See Ibid., 34/100.
21. Schelling, *Introduction*, 231/74.
22. Ibid.
23. Ibid., 230–1/64.
24. Ibid., 202/241.
25. Schelling, *First Outline*, 169/240–241.
26. For these readings, see Friedrich Wilhelm Joseph Schelling, "*Timaeus*" (1794), ed. Hartmut Buchner (Stuttgart-Bad Canstatt: Frommann-Holzboog, 1994); for commentary, see inter alia Manfred Baum, "The Beginnings of Schelling's Philosophy of Nature," in *The Reception of Kant's Critical Philosophy*, ed. Sally Sedgwick (Cambridge: Cambridge University Press, 2000); and John Sallis, "Secluded Nature: The Point of Schelling's Reinscription of the '*Timaeus*'," in *Pli: The Warwick Journal of Philosophy* 8 (1999):71–85.
27. Friedrich Wilhelm Joseph Schelling, *Philosophical Investigations into the Essence of Human Freedom* (1809), trans. Jeff Love and Johannes Schmidt (Albany: State University of New York Press, 2006), 29.
28. See Georg Friedrich Wilhelm Hegel, *Jenaer Systementwürfe III: Naturphilosophie und Philosophie des Geistes (1805–1806)*, ed. Rolf-Peter Horstmann (Hamburg: Meiner, 1987), 160–161.
29. Georg Wilhelm Friedrich Hegel, *Elements of the Philosophy of Right*, trans. Hugh Barr Nisbet (Cambridge: Cambridge University Press, 1991); German version: *Grundlinien der Philosophie des Rechts*, in *Werke in zwanzig Bänden*, ed. Eva Moldenhauer and Karl Markus Michel (Frankfurt: Suhrkamp, 1970), vol. 7:§166, p. 206.
30. Hegel, *Elements*, §165, p. 206.
31. Ibid., §161, p. 200.
32. Georg Wilhelm Friedrich Hegel, *Philosophy of Nature*, trans. A. V. Miller (Oxford: Oxford University Press, 2004); German version: *Enzyklopädie der philosophischen Wissenschaften vol. II*, in *Werke in zwanzig Bänden*, ed. Eva Moldenhauer and Karl Markus Michel (Frankfurt: Suhrkamp, 1970), vol. 9, §368, p. 411.
33. Georg Wilhelm Friedrich Hegel, *Lectures on Natural Right and Political*

Science: The First Philosophy of Right, Heidelberg 1817–1819, with Additions from the Lectures of 1818–1819, trans. J. Michael Stewart and Peter C. Hodgson (Berkeley: University of California Press, 1996), §75, p. 139.
34. So Hegel always takes it. See Georg Wilhelm Friedrich Hegel, *Philosophy of Mind*, trans. William Wallace. (Oxford: Clarendon Press, 1971); German version: *Enzyklopädie der philosophischen Wissenschaften vol. III*, in *Werke in zwanzig Bänden*, ed. Eva Moldenhauer and Karl Markus Michel (Frankfurt: Suhrkamp, 1970), vol. 10:§381A, pp. 9–10.
35. Hegel, *Philosophy of Nature*, §368A, p. 412.
36. Ibid., 413.
37. Ibid., §355A, p. 376.
38. Ibid., §368A, p. 413.
39. Ibid.
40. Ibid., §246A, p. 9.
41. Ibid., §381A, p. 13.
42. Ibid., §389A, p. 10.
43. Ibid., §246A, p. 13.
44. Ibid., §368A, p. 413.
45. Ibid., 413–414.
46. Ibid., §392–§393, pp. 36–45.
47. Ibid., §390–§391, pp. 34–35.
48. Ibid., §393, p. 40.
49. Ibid., §393A, p. 41.
50. As Sara Figal notes in her chapter in this volume, Hegel further distinguishes within the Caucasian race between the inferior Asiatics and the superior Europeans, thus moving the Caucasian character away from the actual Caucasus.
51. Hegel, *Philosophy of Nature*, §370R, p. 416.
52. Peter Hanns Reill, *Vitalizing Nature in the Enlightenment* (Chicago: University of Chicago Press, 2005), 220–235.
53. Philippe Huneman, "The Hermeneutic Turn in Philosophy of Nature in the Nineteenth Century," in *The Edinburgh Critical History of Philosophy*, vol. 5 of *The Nineteenth Century*, ed. Alison Stone (Edinburgh: Edinburgh University Press, 2011), 82.
54. Reill, *Vitalizing Nature*, 234–235.
55. Schelling, *First Outline*, 185/263.
56. Ibid., 184/262.
57. Ibid., 158/230.

58. In Hegel, *Philosophy of Nature*, §368A.
59. Reill, *Vitalizing Nature*, 234–235; and his chapter in this volume.
60. For instance, while Friedrich Schlegel celebrates the chemical mixing of the sexes in his *Athenaeum Fragments*, he says that women can only intuit the Infinite but have no sense for abstractions (1991:102, 30)—that is, women are intuitive, men intellectual.

CONTRIBUTORS

ROBERT BERNASCONI is the Edwin Erie Sparks Professor of Philosophy at Pennsylvania State University. He has published widely in recent continental philosophy and social and political philosophy, on Heidegger, Hegel, Levinas, and critical race theory. He is the author of *How to Read Sartre* (New York: W.W. Norton & Co, 2007), the co-editor, with Sybil Cook, of *Race and Racism in Continental Philosophy* (Bloomington: Indiana University Press, 2003), and an editor, with Katryn Gines and Paul C. Taylor, of the journal *Critical Philosophy of Race*, University Park at Pennsylvania State University Press.

PENELOPE DEUTSCHER is a professor in the Department of Philosophy at Northwestern University. Her publications include *The Philosophy of Simone de Beauvoir: Ambiguity, Conversion, Resistance* (Cambridge: Cambridge University Press, 2008), *How to Read Derrida* (New York: W.W. Norton & Co, 2005), *A Politics of Impossible Difference: The Later Work of Luce Irigaray* (Cornell University Press, 2002), and *Yielding Gender: Feminism, Deconstruction and the History of Philosophy* (New York and London: Routledge, 1997).

SARA FIGAL works as an independent scholar in Nashville, Tennessee. Her area of specialization is the literature and scientific culture of the long eighteenth century. She has received numerous awards for her work, including the Goethe Society of North America's Essay Prize in 2009 and the Max Kade Award for the Best Article in the *German Quarterly* in 2006. She is the author of *Heredity, Race, and the Birth of the Modern* (London and New York: Routledge, 2008 and 2010) and co-author of *The German Invention of Race* (Albany: State University of New York Press, 2006 and 2007). She has published articles on eighteenth-century literature, science, and culture in journals including *Lessing Yearbook*, *Eighteenth-Century Studies*, *German Quarterly*, and *Women in German Yearbook*, and in various essay collections. She has served on the editorial board of the *German Quarterly* and on the National Steering Committee for Women in German.

JOCELYN HOLLAND is currently associate professor in the German program at the University of California, Santa Barbara. Her areas of research include intersections in the literature and science of the eighteenth and nineteenth centuries. Recent books include *German Romanticism and Science: The Procreative Poetics of Goethe, Novalis, and Ritter* (London and New York: Routledge, 2009) and a bilingual edition of the Romantic scientist Johann Wilhelm Ritter titled *Key Texts of Johann Wilhelm Ritter (1776–1810) on the Science and Art of Nature* (Leiden: Brill, 2010).

SUSANNE LETTOW is currently adjunct professor (*Privatdozentin*) at the Institute of Philosophy of the University of Paderborn, Germany. Her areas of research include social and political philosophy, history and philosophy of the life sciences, and feminist philosophy. She is the author of *The Power of Care: Philosophical Articulations of Gender Relations in Heidegger's "Being and Time"* (German; Tübingen: edition diskord, 2001) and of *Biophilosophies: Science, Technology and Gender in Contemporary Philosophical Discourse* (German; Frankfurt a. Main and New York: Campus, 2011).

RENATO G. MAZZOLINI is a professor of the history of science at the Faculty of Sociology of Trento University Italy. He has been a Wellcome Research Fellow at Oxford University, a fellow of the Humboldt-Stiftung at Göttingen, and fellow of the Woodrow Wilson International Center for Scholars in Washington. He specializes in the history of physiology, microscopy, and physical anthropology for the period 1640 to 1850, and he has published extensively in these research areas. He has also edited the unpublished correspondence of several physiologists and natural historians of the eighteenth and early nineteenth century. He is a member of the German Academy of Sciences Leopoldina and of the Akademie der Wissenschaften zu Göttingen.

STAFFAN MÜLLER-WILLE is senior lecturer at the University of Exeter and associated there with the ESRC Centre for Genomics in Society and the Centre for Medical History. He received his PhD in philosophy from the University of Bielefeld, Germany, and has previously worked for the German

Hygiene Museum, Dresden, and the Max-Planck-Institute for the History of Science, Berlin. He has published extensively on the history of natural history, heredity, and genetics. Among the most recent publications is the book *A Cultural History of Heredity* (Chicago: University of Chicago Press, 2012), co-authored with Hans-Jörg Rheinberger. Another project he is working on deals with the ways in which Carl Linnaeus assembled and processed information on plants and their medical virtues. This project is currently funded by the Wellcome Trust.

PETER HANNS REILL is distinguished professor of history emeritus at the University of California, Los Angeles, and emeritus director of UCLA's Center for Seventeenth and Eighteenth-Century Studies. He is a senior fellow at the School of International and Public Affairs at Florida International University (Miami). He specializes in the intellectual history of eighteenth-century Europe with an emphasis on Germany, France, and Great Britain. He has focused on the intersections between science and culture and is also working on a study of the widespread appeal of hermetic thought in the last half of the eighteenth century. Among the most recent publications is the book *Vitalizing Nature in the Enlightenment* (Berkeley, Los Angeles, London: University of California Press, 2005). He has also published *The German Enlightenment and the Rise of Historicism* (Berkeley, Los Angeles, London: University of California Press, 1975) and has edited, among other books, with David Philip Miller, *Visions of Empire: Voyages, Botany, and Representations of Nature* (Cambridge: Cambridge University Press, 1996). He has held fellowships from the John Simon Guggenheim Foundation, the Fulbright-Hayes Program, the Wissenschaftskolleg in Berlin, the Center for Advanced Studies (Munich), the Herzog August Bibliothek in Wolfenbüttel and the Max-Planck-Institut für Geschichte in Göttingen. He is a past president of the American Society for Eighteenth-Century Studies and has been elected a member of the Göttingen Academy of Sciences.

JOAN STEIGERWALD is an associate professor in the science and technology studies program and Department of Humanities, and the graduate program in social and political thought. She is currently completing a book project *Organic Vitality in Germany around 1800*, which examines understandings of organic vitality that developed at the turn of the nineteenth century through new experimental practices, instruments of judgment, and forms of figurative representation. She has published numerous articles on German Idealism and

Romanticism and the history of the life sciences. Her wider research interests are in the intersections between science and technology studies, continental philosophy and contemporary theory.

ALISON STONE is professor of European philosophy at Lancaster University, United Kingdom. She works in feminist philosophy and theory, with particular interests in French feminism and debates around embodiment, essentialism, sexual difference, and sex and gender. She also works in continental philosophy, especially German Idealism and Romanticism, Marxism, and the Frankfurt School. Her books include *Petrified Intelligence: Nature in Hegel's Philosophy* (Albany: State University of New York Press, 2004), *Luce Irigaray and the Philosophy of Sexual Difference* (Cambridge: Cambridge University Press, 2006), and *Feminism, Psychoanalysis and Maternal Subjectivity* (London and New York: Routledge, 2011). She has also edited *The Edinburgh Critical History of Philosophy, Volume 5: The Nineteenth Century* (Edinburgh: Edinburgh University Press, 2011).

FLORENCE VIENNE is a researcher at the Department for the History of Pharmacy and Science of the Technische Universität Braunschweig, Germany, and the coordinator of the scientific network *Economies of Reproduction: Interdisciplinary Research on the Past and Present of Human Reproduction, 1750–2010* (funded by the German Research Foundation). Her main research and teaching areas are history of science/biomedicine (nineteenth and twentieth century), and gender studies. She is currently completing the book *From Spermatic Animalcules to Sperm Technologies: A Cultural History of Semen, 1749–1945*. She is the author of *Une science de la peur: la démographie avant et après 1933* (Frankfurt a. Main: Peter Lang, 2006), and the co-editor, with Christina Brandt, of *The Human as Epistemic Object: Practices of the Human Sciences in the 20th Century* (German; Berlin: Kadmos, 2008).

INDEX

A

abolition, 7, 12, 48, 145, 147–149, 193, 194, 202
Absolute, 33–35, 70
acoustics, 10, 85, 97, 102n22; Johann Wilhelm Ritter on, 89–94. *See also* organism
aesthetics: in Blumenbach's species-description, 178–179; of listening, 92; and race, 164, 166, 172; and skin color, 138, 144, 151
Adam and Eve, 134 135, 138, 140, 270. *See also* creation
African, Africans, 11, 137–138, 187, 190–191, 197, 202: in contrast to Caucasian in race theory, 174, 179; Ethiopian as term for, 137; in Hegel, 272; in Kant, 223–224; skin color in (European) scientific inquiry, 137–152, 226
albinism, 138–140, 226–229
America, Americas, 66, 134
Americans (Native), Amerindians: crossbreeding of, 138, 245; and the New World discovery, 134; in scientific classification, 143, 164, 224–226; women, 189
analogy, 187–205, 206n4, 202n4
anatomy, 109, 112, 188, 206n4; comparative, 4, 116–117, 164; of Galen 135, 137–140, 146; reproductive, 268–271, 275–276 (Vesalius)
androgyny, 67–72, 276–277
animal, animality, 3, 12, 21, 27, 29, 32, 35–36, 75–76, 78, 80, 84, 97, 106, 115–118, 121, 135–136, 138–139, 187, 189–204, 221, 238, 240, 243, 247, 263, 267–273; spermatic, 45–57, 75. *See also* plant; polyp
animalcules, animalculist, 4, 45–47, 74–76
anthropology, 178, 218; philosophical, 219; physical, 4, 131–132, 164, 166, 230
ape (monkey), 120, 135–136, 149, 210n38
Aristotle, 35, 47, 71, 221–222
Aryan race, 164, 177
Aryanism, 177
Asia, Asians (Asiatics), 144, 151, 161n52, 174,176–177, 224–226, 272

B

Baer, Karl Ernst von, 85, 95–97
beauty, 147; Caucasian (Circassian) 69–70, 163–171, 177–182; Georgian, 179; improving and

beauty (*continued*)
  transmitting, 171–173, 175–176; racial, 144
being, 2, 22, 24, 28, 33, 80, 88, 269; order of, 24, 80
Bernier, François, 166, 169–170
Bible, 7, 84, 134, 140–141, 175, 238
Bildungstrieb (formative drive), 4–5, 13, 29, 50, 69, 115, 237–248, 251. *See also* Blumenbach
biology, 2, 5, 9–14, 46, 81, 85, 98, 217–219, 275; biopolitics, 6, 8–9, 24, 37, 152, 180; history of, 53, 56–57; Treviranus, 105–123
blacks, blackness, 136–151, 171, 179, 202, 217, 223, 225, 250; mulatto, 227, 238–245
blood, 11, 41n57, 171–173, 176, 180–181
Blumenbach, Johann Friedrich: formative drive (*Bildungstrieb*), 30–33, 50, 55, 237–244, 247–248; generation theory, 45–46, 50–52, 55, 67, 69, 116–117; hybridity (interbreeding) in, 245–247; and invention of race, 164, 177–179; against preformation, 249; reproductive theory, 83; *Über den Bildungstrieb und das Zeugungsgeschäfte*, 29, 50, 115, 245. *See also* aesthetics
breeding, 163, 167, 180, 227; crossbreeding (race-mixing), 2, 138, 171–173, 178; inbreeding, 180, 227; interbreeding, 121, 246; improvement, 179–180
Buffon, Georges-Louis Leclerc, Comte de, 23; anthropology as natural history in, 132, 142; influence on Mary Wollstonecraft, 195, 204, 206; and Kant, 225, 238–239; generation in, 25, 67, 112, 115–116; reproduction in, 28–29, 37; *Histoire Naturelle*, 29, 46; semen in, 46–57

C

Canguilhem, Georges, 54–55
Carus, Carl Gustav, 77–80, 82n33, 276
castas, 225–26
Caucasian, Caucasians, 7, 11, 163–167, 171, 173–182, 272; Caucasus, in Blumenbach, 164, 177–179; in Meiners, 177–178
cell theory, 53–56, 139
Chardin, John, 11, 166–74, 183n12
chemistry, 83, 85, 108–111, 262–263; Johann Wilhelm Ritter on, 90, 93–98
Chladni, Ernst, 91–92
Christian, Christianity, 22, 66, 133, 146, 150–151, 165, 170, 174, 177, 212n43; origin narrative, 101n22, 134, 270
Circassian, Circassians, 7, 11, 163–182
civilization, 132, 152, 198, 224
climate, 115–116, 118, 120, 140, 142, 149, 224, 238–240, 248, 257n47
colonialism, colonization, 22, 131, 134, 181, 225
cosmology, 9, 23, 25, 29, 31–32, 36–37; Plato, 265
creation, 21, 25–26, 27, 29 32, 52, 69, 71, 74, 80, 89, 115; Christian origin narrative, 2, 22, 23, 26, 32, 45–46, 53, 66, 101n22, 134; *creatio ex nihilo*, 24. *See also* Adam and Eve; Noah
Cuvier, Georges, 119, 136, 144

D

Darwin, Charles, 154n11, 218, 228–230
death, 36, 57, 88, 90, 95–97, 113, 119–120

degeneration, 8, 10, 107, 114–122, 174, 180, 230, 245
Descartes, René, 135
diremption (*Entzweiung*), 34, 275
disease, 110, 180, 207, 226; hereditary, 4, 12, 218, 219–223, 227, 229
dualism, 9, 32, 36–37, 51–52, 55, 57

E
egg (Lat. *ovum*), 25, 28, 47, 52, 56–57, 240, 242
emancipation. *See* woman, women: women's rights
embryo (fetus), 45, 47–48, 52, 56, 74, 80, 220–221, 242
Enlightenment (*Aufklärung*), 10, 54–55, 65–77, 80–81, 224
environment, 68, 106, 115, 122–123, 140, 188–189, 219, 223, 229, 238–239, 250–251, 265, 272–273
epigenesis, 8, 12, 51–52, 66, 74, 80, 237, 241, 243–244, 249
equality, 3–4, 8, 57, 65, 148, 192
Equiano, Olaudah, 12, 187, 192–195, 203–204, 209n32, 210n36,
ethnology, 133. *See also* anthropology
Europe, Europeans, 7, 11–12, 22, 66, 80, 189, 193–194, 197–203, 223–224, 272; colonies, 226; identity, 131–152 (somatic), 163–182; languages, 219

F
Fabricius, Johann Christian, 148–149, 210n38; pupil of Carl Linnaeus, 148
family, 22, 84, 224, 275
female, 163, 166–171, 179–181, 274–275, 191; female as such, 77–81; and male, 5, 33–35, 48–49, 52–57, 65–77, 135, 221, 259–262, 265–277. *See also* history; plant

fertilization, 28, 50, 56–57
Fichte, Johann Gottlieb, 10, 33, 83, 85, 88–89, 95–97
forces, 9, 50, 69, 106, 174, 228–229, 244; attraction/repulsion, 206; Kielmeyer's model of forces, 36; male/female, 69–72; production/inhibition, 13, 259–265, 275–276; propulsion, 108–109; reproductive force, 30–31, 34, 36, 41n57; vital forces, 30
formative drive. *See Bildungstrieb*; Blumenbach
Forster, Georg, 68, 118, 164, 241, 249–250
Fortpflanzung (propagation), 10, 21, 24, 30, 38n13, 83–98, 100n17. *See also* Herder
Foucault, Michel, 2, 65, 227
Frank, Johann Peter, 180–181, 186n53
French Revolution, 4, 10, 54, 73, 193, 227, 275

G
Galen, 48, 135, 221
Garbo, Dino del, 220–222, 227
gender, 2–14, 14n4, 22, 28, 37, 57, 65–81, 171, 173, 265–267; complementarity, 35; *Geschlecht*, 4, 22, 38n5, 274; hierarchy, 51–55; politics, 7, 12
genealogy, 3–4, 8, 21–26, 36–37, 92. *See also* generation
generation (Lat. *generatio*), 2–14, 21, 23, 24–37, 45–57, 65–81, 83–84, 88–89, 105, 112, 114–122, 221, 228, 237, 240–243; spontaneous, 52, 55, 74
geography, 114, 117–118, 164, 175, 244; geographical zones, 118
geology, 114, 117, 119–120

Georgians, 163, 165–168, 172–179
germ, 2, 12–14, 30, 45, 50–52, 237, 241–243, 248–250. *See also* seeds
German Idealism, 41n51, 276
Girtanner, Christoph, 12–13, 111, 243–247, 251, 256n46, 257n49, 258n58
Goethe, Johann Wolfgang von, 10, 68, 71, 75, 85–89, 93, 95, 97. *See also* metamorphosis
Görres, Joseph, 9, 23, 32, 34–36

H

Haiti: St. Domingue, 148–49; Haitian Revolution, 159n41
Haller, Albrecht von, 241–242
Hardenberg, Friedrich von (Novalis), 83
Harvey, William, 47, 220
Heeren, Arnold, 151
Hegel, Georg Friedrich Wilhelm, 9, 13, 23, 32, 35–36, 80, 177, 280n50; sexual polarity in, 259–260, 266–277
Henrichs, Hermann, 167–169
Herder, Johann Gottfried, 9–10, 23, 29–34, 37, 41n51, 41n57, 95, 97, 100n16, 101n22, 145, 164, 249; on *Fortpflanzung*, 85–91
heredity (*Vererbung*), 1, 3–4, 7, 11–13, 22, 28, 217–231, 239, 246
Hippocrates, 48; Hippocratic explanation, 67, 222
history, 12, 87, 92, 94, 133, 151, 175–177, 197, 218, 273; as feminine, 69; natural, 3, 5, 22, 26, 33, 45–46, 106–111, 114–123, 132, 136–137, 142–144, 146–147, 164, 167, 187, 204, 224, 227, 237–251, 230; sacred, 134–135. *See also* science: history of
Honegger, Claudia, 3, 65
Hufeland, Christoph Wilhelm, 110–111

Humboldt, Wilhelm von, 9, 68–72, 75, 112, 118–119
human species, 2–3, 21, 30–31, 46–52, 67–72, 77, 80, 87, 109, 115–116, 120, 131–147, 163–167, 173–182, 188–197, 199, 217–218, 222
humanity, humankind (*Menschheit*), 22–24, 37, 66, 75, 88–97, 190, 193–194, 197, 201, 223–231, 238–250, 261–276
humoral pathology, 219–221
hybridity, 48, 75–77, 121, 165, 174, 177, 179, 223, 238, 242–247; half-breed, 3, 240, 246. *See also* Blumenbach

I

individual, individuality, 1, 4, 22–37, 49, 54–57, 67, 71, 74, 84–85, 88–91, 96, 121, 179, 228, 243, 247–249, 259, 264, 271, 273
infusorians, 52–56, 106, 118–119
inheritance, 121, 219–230, 239–241, 245, 250. *See also* heredity
intermarriage (mixed marriage). *See* marriage
Islam, Islamic, 163. *See also* Muslim, Muslims

J

Jacob, François, 21, 53–55
Jefferson, Thomas, 144–145, 157n26

K

Kant, Immanuel, 3–5, 33, 72, 83, 107–108, 110, 122, 144, 260–261; *Critique of Judgment*, 2, 8, 41n48, 72; *Fortpflanzung*, 100n17; on generation, 115 ; heredity, 4, 219, 223–224, 237–251; history of nature, 115–116, 255n30; on race, 3, 12–13, 225 156n20, 164, 237–251

Kielmeyer, Carl Friedrich, 9, 23, 29, 31–32, 34, 36–37, 41n62, 83, 111
kinship, 4, 8, 22–24, 28, 37, 84
Kölliker, Albert, 56–57

L

La Mettrie, Julien Ofray de, 23, 26–27
Lamarck, Jean-Baptiste, 22, 136, 228
Laqueur, Thomas, 65–66, 76, 260, 274
Lavater, Johann Casper, 179, 249
Lavoisier, Antoine, 83, 111
Leclerc, Georges-Louis Marie. *See* Buffon, Georges-Louis Leclerc, Comte de
Leeuwenhoek, Antoni van, 45–47
leucocracy, 151
life, 2, 10, 22, 28, 31, 35–36, 48–49, 53, 56–57, 68–69, 106–123, 189, 226–230, 267; interaction between living forces, 50–51, 69, 86; *Lebenskraft, -kräfte*, 106–123; the one (*Monas*), 24, 80
Linnaeus, Carl, 3, 118, 136, 143–144, 195, 204–205, 211, 217–218, 223, 226, 228. *See also* Fabricius
Lucas, Prosper, 228–229

M

Macaulay, Catherine, 12, 187–188, 190–193, 195–199, 202, 215n87
male, 35–36, 54–57, 69–77, 80, 165–167, 222, 242, 259, 262–266, 268–277
Malpighi, Marcello, 139–140, 143, 154n12
marriage, 8, 17n34, 197; intermarriage, 180, 225
Marx, Karl, 227
masculinity, 35, 67–72; male principle, 36, 69, 72, 74–78, 276; pure masculinity, 36, 74–80
materialism, materialist, 26
matter, 13, 29, 35, 47–56, 69, 74–75, 80, 107, 108, 110, 112, 259, 262, 274; concept and, 266–271; material, 32, 48–50, 52, 74, 83, 89, 92, 94, 107, 113–115, 122, 259, 265, 270–276; organic, 94–95, 106, 112; *lebensfähige Materie* (viable matter), 113–123
Maupertuis, Pierre-Louis Moreau de, 2, 4, 23, 26–29, 37, 48, 53, 59n19, 226–228, 239
medical police (*Polizeywissenschaft*), 173, 180
medicine, 6, 106, 109, 111, 218–220; biomedicine, 12–13
Meiners, Christoph, 147, 164, 175–178
Mercado, Luis de, 221–222
metamorphosis, 85–86, 89, 97
Metzger, Johann, 249, 254n29
Mongolia, Mongols, 151, 164, 175, 180, 272–273
monogenesis, monogenists, 2, 132, 157n27, 238, 241, 250, 254n29
monster, monstrosities, 28, 51, 121, 227, 246. *See also* degeneration
Muslim, Muslims, 11, 165, 171, 174, 177. *See also* Islam

N

natural law, 101n17, 222–224, 228
natural history, 3, 5, 22, 33, 45–46, 106–111, 115–117, 132, 136–143, 147, 164, 187, 227–251, 255n30. *See also* Kant
Naturphilosophie, 5, 6, 9, 32–37, 57, 69, 72–81, 84, 113, 260, 262, 274–277. *See also* Oken; Schelling
Needham, Joseph, 106, 112, 125n24
Negro, 144–147, 191, 226–227, 241, 272, 273
Newton, Newtonian science: atomism, 262; gravitational force, 30, 50; mechanics, 53, 108; physics, 261

Noah, Flood narrative, 7, 135, 140, 160n48, 175
nobility, 173, 227
Nogay Tartars, 178–179
nutrition, 5, 28–30, 49–50, 114, 117, 219, 223, 242

O

offspring, 94–95, 172, 178–179, 218, 220, 222–225, 228, 264–271
Oken, Lorenz, 9–10, 36, 46, 51–57, 61n60, 72–78, 261, 276
organic molecules, 9, 29, 45–57, 59n25
organism, 22, 29–35, 45, 47–56, 74, 76, 78, 83–84, 86, 105–122, 229–230, 239, 244, 261, 268; tone as, 91–93
Orientalism, Orientalist, 11, 138, 168, 171, 174, 176, 213n71
origin, 24, 27–28, 32, 34–37, 43n98, 48, 51, 55–56, 74–75, 80, 97, 107, 119–120, 134–135, 138, 163–165, 174, 176–182, 224, 239, 248, 255n34
Ottoman Empire, 163–165, 169, 174, 181
ovaries/testes, 47–48, 268–271

P

parents, 24, 29, 48, 50–51, 90, 179, 202, 221–223, 225, 240, 242, 245–246, 267–268
Pechlin, Johannes, 139, 147
Persia, Persians, 166, 170–174, 181
philosophy, 2, 5–6, 10–14, 54, 80–81, 107–110; of nature, 5, 33–35, 66, 84, 114, 137, 259–261, 266–267, 269–277; as masculine, 69; transcendental, 33. See also Naturphilosophie
phylum (stem, *Stamm*), 3, 25, 240–241, 248

physiology, 25, 34, 56–57, 106, 108–111, 114, 117, 122, 137, 188, 212n47, 263, 284
plant, 10, 21, 29, 35, 49, 51, 75–78, 117–118, 121, 195, 206; as female, 54, 78, 84–89, 93, 97, 204, 211n41, 215n91, 242–243, 264, 276; plant-animals/animal-plants (*Phytozoa*), 10, 106, 117
Plato, 71, 265, 269
polygenesis, 132, 241, 250
polyp, 50, 74–76, 276; Abraham Trembley's experiments, 23, 25–30, 40n31; reproduction of, 23, 27; Lorenz Oken, 36, 276; male principle, 77
predispositions (*Anlagen*), 115, 238, 241, 244, 248
preexistence theory, 4, 45–47, 51. See also preformation
preformation, preformationism, 2–4, 8, 12, 40n31, 66, 74, 237–38, 241, 249; animalculism, 4, 45, 47; critique of, 26–27, 51, 242–44; ovism, 47, 51
Prichard, James Cowles, 144, 250
procreation (*Zeugung*), 1–4, 8–10, 21, 29, 35, 222, 271; discourse, 83–98, 99n7. See also Fortpflanzung
production (Lat. *productio*), 13, 21–24, 27–28, 34, 84, 91, 93–94, 219, 230; Schelling's polarization (production/inhibition), 260–66. See also reproduction
progress, 69, 72, 107, 133, 152, 191, 224, 230
Prussia, 73; academic reform in, 6
purpose (*Zweck*), 2, 87–88, 97, 243, 268, 275. See also teleology

R

race, 1–14, 28, 37, 72, 75, 80,

115–116, 121, 131–132, 144–147, 151–152, 163–182, 188, 191, 194, 202, 217–231; Blumenbach's concept of, 245; Girtanner on, 246–247; Hegel on, 272–273; Kantian notion of, 237–241, 244, 246, 249–251; Schelling on, 248–249
racism, 6, 11, 132, 145
recapitulation, 30–32
regeneration, 2, 5, 8, 30, 50
Reil, Johann Christian, 108, 110–112
reproduction, 1–4, 7–11, 13, 21–37, 56, 84–85, 92, 94, 112, 121, 136, 219, 224, 245; Blumenbach on, 242; Buffon on, 46–50, 57, 78; and deviation, 228, 230; Hegelian, 259, 266–267, 270, 273; Kantian, 115, 241; and racial traits, 172–173; Schelling on, 263–264
Ritter, Johann Wilhelm, 10, 83, 85, 89–98. *See also* acoustics; chemistry
Romanticism, 65, 81, 83

S

Saint Vincent, Bory de, 144, 147
savage, savages, 13, 151, 189–195, 198–202, 208n9, 208n10, 210n36; noble, 133
Schelling, Friedrich Wilhelm Joseph, 5, 9, 12–13, 54, 97, 243–244; *Erster Entwurf eines Systems der Naturphilosophie* (First Outline), 84, 113–114; influence of Kant and Blumenbach, 247–249, 251; *Naturphilosophie*, 32–36, 73–75; sexual polarity in, 259–266, 269, 271, 273–277
Schiller, Johann Christoph Friedrich von, 68, 72; *Über die Ästhetische Erziehung des Menschen*, 72
Schlegel, Friedrich, 71, 73, 83
science, 65, 79, 95, 147, 152, 164, 218, 237, 246, 261–263, 269; history of, 1, 12, 105, 237, 260; life sciences, 1–14, 30, 33, 45–46, 83, 105–110, 114, 123, 230–231
seeds (*Keime*), 80, 87, 96, 220–222, 237, 241–243, 271. *See also* germs
sex (gender), 2–3, 8, 10, 52, 167, 243; sexual difference, 4, 7, 32–37, 55–57, 66–81, 189–191, 195, 213n61; sexual polarity, 13, 14n1, 38n5, 22–24, 27–28, 259–277; two-sex model, 5, 47–49, 65, 85, 207n4
skin color, 7, 11, 28, 136–149, 164, 175, 178, 217, 223–226, 239–240, 246, 257n49
slavery, 7, 11–12, 145–152, 171, 191–196, 198, 202, 204–205; slave trade, 147–148, 169, 172
Smith, Samuel Stanhope, 187–189, 193, 195, 202, 208n7
soul, 13, 196, 272–274; animal, 26, 40n31; world, 114, 262. *See also* Schelling
South America, South Americans, 174, 225–226. *See also* America; *castas*
species (*Gattung*, *Art*), 3, 13, 22, 24–26, 28–37, 66–69, 78–79, 85–89, 93, 95–97, 100n17, 115–121, 134–138, 143–146, 49, 163–166, 173, 176, 178–180, 188–191, 218, 223–224, 227–229, 238–248, 264, 266–268, 273, 276
sperm, spermatic animals (semen), 4, 9, 45–57, 75–76, 140, 220, 268, 271, 240–242
spirit, spiritual, 35–36, 70, 79, 114, 135, 270, 272; spiritual ladder (Goethe), 86–97
spontaneous generation. *See* generation
Stamm. *See* phylum
subjectivity, 33, 65, 181, 260, 262, 265, 271

T

Taylor, Barbara, 187, 192–194, 209n31
teleology, 41n48, 88–89, 100n17, 178, 244
testes. *See* ovaries
theology, 6, 84, 133, 137–138, 140–141, 146, 221
time, temporality, 9, 25–32, 34–37, 75, 86, 90, 93, 31–32, 37; temporalization, 1–2, 22–23, 34
travel literature, 7, 11, 118, 133–134, 137, 141, 163–182, 225
Trembley, Abraham, 23, 25–30. *See also* polyp
Treviranus, Gottfried Reinhold, 10–11, 105–123
Turkey. *See* Ottoman Empire

U

ugliness, 7, 172–173, 175, 178–179, 81

V

Vandermonde, Charles Augustin, 79–80
vitalism, 4, 30, 54, 67–77, 80
Voltaire, 169–170

W

whites, whiteness, 7, 11, 80, 138–140, 143–151, 163–171, 175–182, 217, 223, 225–227, 241, 248, 250
Wollstonecraft, Mary, 8, 12, 68, 187–205. *See also* abolition
woman, women, 7, 9–10, 47, 70–71, 78–80, 163–182, 189–192, 275–277; contribution to generation (procreation), 52, 54, 77; women's rights, 4, 8, 12, 68, 187, 192–205. *See also* Circassian

Z

Zeugung, Zeugungskraft. *See* procreation
zoophytes, 106, 117–121